Civilizing Climate

Civilizing Climate

Social Responses to Climate Change in the Ancient Near East

ARLENE MILLER ROSEN

A Division of
ROWMAN & LITTLEFIELD PUBLISHERS, INC.
Lanham • New York • Toronto • Plymouth, UK

ALTAMIRA PRESS
A Division of Rowman & Littlefield Publishers, Inc.
A wholly owned subsidiary of The Rowman & Littlefield Publishing Group, Inc.
4501 Forbes Boulevard, Suite 200
Lanham, MD 20706
www.altamirapress.com

Estover Road
Plymouth PL6 7PY
United Kingdom

British Library Cataloguing in Publication Information Available

Library of Congress Cataloging-in-Publication Data

Rosen, Arlene Miller.
  Civilizing climate : social responses to climate change in the ancient near East /
Arlene Miller Rosen.
     p. cm.
  Includes bibliographical references and index.
  ISBN-13: 978-0-7591-0493-8 (cloth : alk. paper)
  ISBN-10: 0-7591-0493-X (cloth : alk. paper)
  ISBN-13: 978-0-7591-0494-5 (pbk. : alk. paper)
  ISBN-10: 0-7591-0494-8 (pbk. : alk. paper)
  1. Environmental archaeology—Middle East. 2. Paleoecology—Middle East.
3. Climatic changes—Middle East. 4. Excavations (Archaeology)—Middle East.
5. Middle East—Antiquities. I. Title.

  CC81.R67 2007
  930.1—dc22
                                                               2006021682

For Steve, Yaniv, and Boaz

Contents

Figures and Tables

FIGURES

TABLES

Acknowledgments

I am very grateful to my former instructors, Karl Butzer and Fekri Hassan, who were ultimately the underlying inspiration for this work. The project that culminated in this volume would never have gotten off the ground without the initial support of an Independent Scholar's Fellowship from the National Endowment for the Humanities. With this stipend I was able to impose on the patience and resources of the good people of the Cotsen Institute of Archaeology, University of California, Los Angeles, in my early stages of writing. I owe a debt of gratitude to my friends and colleagues at that institute who warmly hosted my family and myself in that intellectually stimulating year. I also would like to acknowledge the support of a Leverhulme Foundation grant to Neil Roberts, Warren Eastwood, and myself, for investigation into the ecological footprint of Epipaleolithic and Neolithic populations in the Near East. This research helped contribute to some of the information and insights in this book.

Over the course of conducting research and writing this volume I have had the privilege of discussing various aspects of this study with a number of my colleagues including Oren Ackermann, Eleni Asouti, Diana Beatty, Sue Colledge, Carlos Cordova, Peter Fabian, Dorian Fuller, Baruch Halpern, Arie Issar, Michael Jasmin, Emma Jenkins, Li Liu, Joseph Patrich, Mark Smith, Hans von Storch, Kai Wirtz, and Henry Wright, among others. I also want to thank Mitch Allen and Peter Ucko, who supported the writing of this volume. My most sincere thanks go to Ofer Bar-Yosef, Neil Roberts, and Ken Thomas, who took time out of their busy academic schedules to read over and comment on some of the chapters of this book. Their help was invaluable, and I learned a great deal from them. I especially thank Diana Beatty, who volunteered her valuable time to help with the preparation of illustrations for this volume. I am very grateful to these friends and colleagues for their input, although any errors of omission, fact, or judgment are entirely my own. My husband, Steve Rosen, was a beacon of support to me while I was writing this work. In addition to providing me with his typically insightful comments and critique, he was a constant source of encouragement and inspiration. I also have a deep debt of gratitude to my parents, Lilyan and Leonard Miller, for instilling in me their love of nature and interest in the natural world, and last but certainly not least I want to express my sincere thanks to my sons, Yaniv and Boaz, who have helped me in fieldwork, provided comic relief, kept me in the *real* world, and have given me a small glimpse into the future of humanity.

1

Holocene Climate and Society: Civilization and Climate Revisited

In 1903, the geologist-archaeologist Raphael Pumpelly led an archaeological expedition to the forbiddingly arid lands of Turkmenistan in Central Asia. There, at the oasis site of Anau, he found the remains of a prosperous Neolithic farming society dating to around 4500 BC (see figure 1.1). The enigma of this once thriving town in what is today a poorly watered desertic region appeared to serve as a silent testimony to the devastating impact of past climatic change on human societies (Pumpelly 1908). It led Pumpelly to develop the foundations for the idea of humans, plants, and animals converging in rich localities as resources began to diminish in an increasingly degrading environment. This concept later came to be known as the "oasis theory." Pumpelly was joined in this expedition by Ellsworth Huntington, an American geographer. Huntington was profoundly influenced by his Asian fieldwork experience and later wrote a landmark book entitled *Civilization and Climate* (1915, rev. ed. 1924). This volume was the defining statement of a position that since has been labeled "climatic determinism." Although extreme in its orientation, the book outlined a philosophy that implicated a climatic and natural environmental cause for many characteristics of human societies and culture change. It was a book that had a notable influence in its time, and together with the work of Pumpelly inspired great anthropological thinkers such as V. Gordon Childe, who depended heavily on climatic deterministic principles in developing the oasis theory for his models of agricultural origins (*New Light on the Most Ancient East* [1929, rev. ed. 1952]).

1

FIGURE 1.1
Raphael Pumpelly's Expedition to Anau (1904).

Since the 1960s, with the introduction to archaeology of concepts such as cultural ecology and systems theory, many archaeologists began to realize that monocausal explanations for major social changes were far too simplistic and often fraught with too many unanswered questions and enigmas. There was an acknowledged need for in-depth research into the interactions of the many segments of a given social system with an equally complex and changing environment (Butzer 1982; Flannery 1972; Rosen and Rosen 2001). Such research was carried out within a framework of processual archaeology with its guiding principles of seemingly objective scientific methods. Currently, in the intellectual climate of the early twenty-first century, environmental archaeology can also benefit from aspects of postprocessual enquiry by adding a more humanistic dimension to the relationship between people and their environment.

Archaeologists now understand that human perceptions of nature, environment, and climate change are very much a key to how societies adjust to the impact of environmental change (Crumley 2001; McIntosh, Tainter, and McIntosh 2000b; A. Rosen 1995). Intellectual orientations such as historical and political ecologies (Crumley 1994; Greenberg and Park 1994; Peet and Watts 1994) suggest that human responses to environmental stress may also be strongly influenced by conflicts in goals and agendas set out by ruling classes versus those of other segments of societies (McIntosh, Tainter, and McIntosh 2000a). This message has a resonance with our own current struggles to come to grips with the concept of global warming and the prospects of secular climatic change and its impact on the modern world.

Given this intricate latticework of multifaceted interactions between humans and their changing environments, this volume attempts to gain a more in-depth understanding of climate change and human social responses from the Terminal Pleistocene through the Late Holocene, a period of time dating from about 13,000 to 1500 years ago. I take the Near Eastern Levant as an example of how climate change impacted different types of communities and segments of communities from hunter-gatherers to early complex societies and empires, and how these societies responded to such changes. This is a period of time that witnessed the most profound transformations in human societies since humanity first appeared on the scene some two million years ago (Roberts 1998). Much of the debate about climate change and its impact on societies in the Near East centers around three important time periods. One of these is the Terminal Pleistocene Younger Dryas (YD) event and its impact on the hunter-gatherer communities of the Southern Levant, with some researchers claiming that the YD was the catalyst that pushed the complex hunter-gathering Natufian societies into shifting to incipient agriculture. The second is the climatic event at 4000 BP that brought widespread drought to the Near East and ushered in the much drier Late Holocene after a moist Middle Holocene. This episode has been blamed for extensive abandonments of Early Bronze Age communities around the Near East. Finally, the Roman/ Byzantine incursions into the arid desert zones of the Near East have been widely attributed to climatic ameliorations in the Late Holocene at around 2000 BP.

Before these perspectives on the relationship between climate and society can be explored, it is essential to obtain accurate scientific data from proxy indicators of climatic change throughout the Holocene. In recent years, a number of proceedings have been conducted in which specialists from diverse disciplines presented new bodies of data on environmental changes during the Late Pleistocene and Holocene in the Near East. This information has helped to refine our understanding of climatic change and environmental fluctuations occurring throughout the Near East during this time period (Bar-Yosef and Kra 1994; Dalfes, Kukla, and Weiss 1997; Roberts, Kuzucuoğlu, and Karabıyıkoğlu 1999). However, many of these data are published in primary form and remain unsynthesized, making it difficult to interpret for many scholars concerned with the history, historical geography, and archaeology of this region. The confusion is increased by dating methods and resolutions that differ between bodies of data, and by the apparent conflict between some lines of evidence for specific time periods. Therefore, in the first section of this volume I review these data and methods and present a summary of climatic changes from the Terminal Pleistocene through the Holocene.

The perspective proposed in the subsequent chapters attempts to examine the interactive relationship between ancient Near Eastern societies and their environments, taking into consideration social organization, technology, and political and economic factors, as well as human perceptions of nature and environmental change. This point of view is very much contrary to a commonly held assumption that societies are monolithic bodies that roll and flow with environmental tides, sometimes succumbing when the forces of nature exceed a society's ability to maximize its resources and adapt (Issar and Zohar 2004). The overall theme of this volume emphasizes that there are many ways in which societies can overcome major environmental shifts. Failure to do so indicates a breakdown in one or more of the social and political subsystems. Both natural and social scientists have much to learn from examining the details of this societal change, even more than we have to gain from compiling a simple tally of failed or dramatically altered societies and comparing this against the highs and lows of environmental change. Finally, environment and climate are not seen here as static, but as naturally variable. This variability is taken as the norm rather than the exception, with environmental fluctuations occurring at different orders of magnitude. The environment is treated here as an important actor in the story of the development of Near Eastern society over the past 13,000 years, but not the sole determinant of social change.

THE QUESTION OF SCALE IN ENVIRONMENTAL CHANGE

Climatic change and environmental shifts occur at differing temporal and spatial scales. These fluctuations sometimes require adjustments or adaptation on the part of human communities that inhabit these environments. The topic of scale in environmental studies has been thoroughly covered in Karl Butzer's classic works on human-environmental relationships, *Environment and Archaeology* (1971) and *Archaeology as Human Ecology: Method and Theory for a Contextual Approach* (1982; see also Dincauze 2000). Although I do not attempt to reiterate the subject in great detail here, it is important to clarify some aspects of scale before proceeding with discussions of climate and society. In environmental studies the concept of scale plays a vital role for the appreciation of the spatial extent, temporal range, and magnitude of environmental shifts (see table 1.1). Butzer defines *large-scale* environmental variability as consisting of biome shifts that persist from millennia to tens of millennia, and major alterations in soil and sediment assemblages resulting from new hydrological and geomorphic systems. *Medium-scale* variability occurs at the magnitude of a few centuries or millennia and encompasses fundamental changes in hydrology with readjustments of stream behavior, as well as changes in biomass with the qualitative composition of plant and animal communities persisting while the quantitative composition changes. The *small-scale* variability is outlined as year-to-year anomalies, up to several decades with no change in stream behavior, and seasonal availability of water as well as plant and animal foods (see table 1.2) (Butzer 1982:23–32).

In studies of Holocene climate and society, the concern here is primarily with medium- and small-scale variability. These changes may seem insignificant from a perspective looking back over the course of millennia, but this does not diminish their importance to contemporary indigenous societies. To people living without the benefit of world market systems and international aid programs, small shifts in rainfall (e.g., 50 mm per year in marginal agricultural regions) can make a difference between bumper crops and famine. Some smaller-scale fluctuations may not always be detectable in the proxy environmental record and are sometimes invisible to paleoclimatologists looking for major trends. In writing about the climatic setting for civilizations in the Southern Levant, some researchers have noted there has been no *macro-climatic* change throughout the Holocene. This observation is based on the charcoal and other botanical evidence for the dominance of Mediterranean

Table 1.1. Scales of Environmental Variation (After Butzer 1982:24, Table 2.2)

First order (less than 10 years): Year-to-year oscillations, including the 26-month atmospheric "pulse," the Great Plains dust bowl of 1934–1939, and the Sahel drought of 1971–1974.

Second order (several decades): Short-term anomalies, such as well-defined trends in the instrumental record, including the Arctic warmup of 1900–1940 AD and the dry spell in East Africa 1900–1960 AD.

Third order (several centuries): Long-term anomalies, such as the worldwide "Little Ice Age" of about 1400–1900 AD or the warm European "little optimum" of 1000–1200 AD, of sufficient amplitude to show up in geological records; third-order climatic variations include repeated oscillations during the 10,000 years of the Holocene.

Fourth order (several millennia): Major perturbations, such as severe interruptions within the last interglacial, the stadial-interstadial oscillations of the last glacial, and the warm and often drier millennia between 8000 and 5000 years ago (altithermal, climatic optimum).

Fifth order (several tens of millennia): Major climatic cycles of the order of magnitude of glacials and interglacials, spanning 20,000 to 70,000 years, with eight glacials verified during the last 700,000 years.

Sixth order (several million years): Geological eras, including the durations of ice ages such as the Permocarboniferous (ca. 10–20 million years long, about 290 million years ago) and Pleistocene (formally began 1.8 million years ago, with major cooling evident for 3.5 million years).

Table 1.2.   Models for Scale Changes in Ecosystems (After Butzer 1982:28, Table 2.4)

Small-Scale Variability (First and Second Order)	Medium-Scale Variability (Third and fourth Order)	Large-Scale Variability (Fourth and Fifth Order
Year-to-year anomalies, or cyclic variations, up to several decades.	Dynamic equilibrium, with major perturbations or low-threshold equilibrium shifts lasting a few centuries or millennia.	High-threshold, metastable equilibrium changes, with biome shifts during the course of several centuries but persisting for millennia, even tens of millennia.
Steady-state or dynamic equilibrium.		
Fluctuations in seasonal availability or aperiodic availability of water, primary productivity, and biomass of plant foods.	Fundamental changes in hydrology, productivity, and all categories of biomass.	Hydrological and geomorphic systems include new components, creating different soil and sediment assemblages.
Affects resource levels for macroconsumers and animal biomass; impact greatest in biomes with low predictability.	Shifts in soil-slope balance favor readjustments of stream behavior, with downcutting or alluviation, tangible in geological record.	New ranking of dominants and subdominants in biotic communities, with transformation in biome physiognomy and biochore definition.
No change in stream behavior or biochore definition.	Qualitative composition of biotic communities persists, but quantitative changes affect mosaic structures in general and ecotones in particular (e.g., species number and selected population densities); minor changes in biochore definition.	Geological and biotic discontinuities provide stratigraphic markers tangible over continental areas.

vegetation throughout that time period (e.g., Liphschitz 1986; Liphschitz, Gophna, and Lev-Yadun 1989). Such a claim prioritizes the macroclimate over the important impact of meso- and microscale climatic fluctuations and thus can lead to a basic misunderstanding of the relationships between human societies in the region and their interactions with the environment throughout the course of the Holocene.

Small- and medium-scale environmental factors are of critical importance for understanding the growth, development, and mutual relations of ancient societies in the Near East. This is especially true of the complex societies that developed during the Holocene and that at times existed very close to the limits of fluctuating carrying capacities of the land they inhabited. As I demonstrate in the ensuing chapters, the environment has fluctuated significantly throughout the Late Pleistocene and Holocene. It was characterized by long phases of ameliorated climate with conditions considerably moister than those of the present and was punctuated by episodes of severe dryness. However, even in these moist episodes, periodic drought was always an overriding threat and limiting factor that had to be acknowledged and accounted for throughout the course of cultural development.

Accounting for climatic oscillation on the small scale is critical for determining the agricultural potential of a given region in the Near East. Many archaeologists have overlooked this aspect when attempting to reconstruct the carrying capacities of their study area. It is common for researchers to estimate the agricultural potential of an archaeological territory by determining the present-day average annual rainfall in the region. If it is equal to or greater than 200 mm for barley or 250 mm of rainfall for wheat (Purseglove 1972:292), then it might be assumed that subsistence farming (albeit marginal) can be conducted in a given area. If the average annual rainfall is higher (from 400 mm up), or if there are proxy data indicating a moister climate, then researchers traditionally assume that agricultural production is secure.

The use of annual rainfall averages to estimate farming potential in the Near East is highly misleading. This is well illustrated by an example from a modern-day marginal zone for rainfall agriculture in south central Israel (see figure 1.2). Rainfall data for Kiryat Gat were collected each year between 1961 and 1988 (D. Amiran, n.d.). Today, the average annual rainfall is about 420 mm, which might appear to be a perfectly adequate amount for cultivation of both wheat and barley. However, if we look at the year-by-year records for

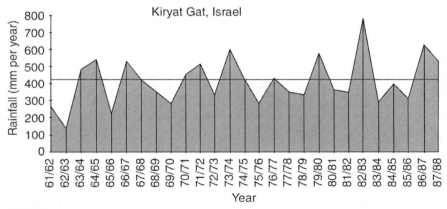

FIGURE 1.2
Modern Rainfall Data over a Twenty-Eight-Year Period for Kiryat Gat, Israel.

rainfall quantities, it is clear that very few years actually receive the average rainfall amounts. Most years are well above or below this measure, with quite a few falling far below, thus resulting in severe water crises. Inhabitants of this region must adjust to the constant threat of drought. A farming strategy designed to support a large settlement cannot depend solely upon dry farming. To do so would be a guarantee of failure. Although individual small-scale subsistence farmers might be able to survive for one or two bad years by utilizing a series of buffering strategies (discussed below), large villages and towns with complex social hierarchies, including the support of a number of nonproducers, would be at the mercy of seemingly random environmental events. Therefore, from the beginning of early farming through the development of early states, it was imperative for dry farming in semiarid regions to go hand in hand with some form of buffer such as floodwater farming, hydraulic manipulation, large-scale storage, and social networks for support in times of famine.

PERCEPTION OF NATURE AND ENVIRONMENTAL CHANGE

Environments change, sometimes catastrophically, but the impact of this change on societies is highly variable. Its effect varies according to technological development, social organization, and perception of this change. Neither a strictly postmodernist view of nature as a purely cultural construct nor the environmentalist perspective of nature as an independent entity will help us to fully explore the relationship between humans and their changing environments. Political ecologists have attempted to reconcile these two fundamentally opposed positions and attain a working method of operation for increased understanding of humans and nature (Greenberg 1994; Peet and Watts 1994; Robbins 2004). Escobar (1999) points out that in any given society there are several different dimensions of environmental perception. These relate both to nature as a cultural construct and to nature as an extracultural essence. In the former, we take action in accordance with our perceptions of nature given to us by our cultural heritage. In the latter we respond to essential nature through "a process of enskillment in practical engagement" (Escobar 1999:9), or—more simply stated—from practical knowledge gained by experience. Following this perspective, this volume takes the stance that societies operate on both the conceptual and practical planes and have the innate potential to adjust to detrimental environmental shifts through either social mechanisms or technological change.

In examining the relationship between environmental change and ancient societies, the climatic determinist perspective has led to a view of social groups as monolithic entities that are impacted uniformly by intense climatic events. They therefore rise, adapt, or fall in a uniform response to these climatic changes (Neev and Emery 1995; H. Weiss et al. 1993). To understand the interactions between societies and their environmental milieu it is critical to remember that communities are segmented in innumerable ways and consist of subgroups and individual actors whose goals and motivations may differ, thus leading to differing responses to environmental factors. Rappaport (1977) indicates this in his discussion of "high-order regulators" versus "low-order regulators." An example of the former group would be the wealthy landowners in a given society, and the latter group would be the peasant farmers. These segments would respond to environmental fluctuations in very different manners that relate to their underlying motivations. A landowner whose main objective is the production of cash crops for a market economy might employ a strategy of maximization. This would entail taking risks to increase the possibility of a bumper crop of a particular commodity. Conversely, subsistence farmers traditionally employ tactics that minimize risks even if this results in lower yields of preferred crops in favor of less preferred but more dependable varieties (Halstead and O'Shea 1989b).

Another example of motivations that differ between social segments comes from the respective responses of Indian peasants versus landowners in times of severe drought. The landowners often benefited from periodic short-term drought because they were able to acquire land and labor at low prices, and so in the long run their wealth and influence increased (Rao 1974). They therefore might have been slower to take protective measures or to encourage technological advances that would have improved agricultural efficiency in new ways. In contrast, the peasant farmers who operated within a subsistence economy needed to employ a variety of measures in the realm of land management, pastoral strategies, and storage systems to protect themselves from drought. This was a strategy aimed at minimizing risk (Halstead and O'Shea 1989b).

Subsistence-level farmers use numerous methods to buffer against the contingency of environmental disasters. These include (a) diversification of cultivated crop species and herded animals to insure a better chance of one or the other species surviving adverse conditions, (b) investment in storage technology, (c) expanding social networks through marriage ties to be able to depend on other communities for support in times of famine, and (d) trading perishable foodstuffs for durable items of exchange that can be converted to food in times of famine (Colson 1979; Gallant 1989; Halstead and O'Shea 1989b). Likewise there are successful responses for surviving times of famine when they occur. Some of these are (a) consumption of seed normally kept for the following planting season, and eating or selling off core herd animals, (b) the secret preparation and consumption of dwindling food supplies, (c) the use of famine foods not normally eaten, (d) the selling of family members, and (e) migration (Dirks 1980).

However, these practical steps and others like them do not always constitute the exclusive solutions to problems of detrimental environmental shifts. Other "nonpractical" solutions are very much a part of the adaptations humans make to their environmental milieu and depend upon the way the society in question views nature in general and environmental change in particular. McIntosh, Tainter, and McIntosh (2000a) propose that societies do not interact directly with their natural environments, rather with their perceptions of these environments. They also point out that to human populations, environmental crises represent a failure to adapt in both the technological and social realms: "The truly interesting questions in such cases involve not just meeting the biophysical challenge but also understanding why human problem solving seems on critical occasions to become ineffective" (McIntosh, Tainter, and McIntosh 2000a:7).

Responses to harsh environmental change are very much related to how a particular community views the reasons for this transformation. Many times, rainfall, drought, heat, storm intensity, and so on are viewed as an integral part of the cosmology, and solutions to problems of adverse weather and climate change must make sense within that framework. For example, if rainfall is a divine gift, then solving the problems related to drought must involve dealings with the supernatural in the form of pleasing the deity responsible. Failure to adjust to environmental stress is as much a social and cosmological problem as an environmental one.

By the same reasoning, times of favorable environments are not de facto signals for automatic growth and economic prosperity. There are often social and political reasons why new resource areas are exploited or production is expanded to take advantage of these favorable environmental changes. Although there are indeed some cases in which environmental shifts coincide with a social collapse or conversely increasing prosperity in a seemingly simple cause-and-effect relationship, when examined closely, the collapse caused by adverse environmental conditions is more often a function of innate fissures already existing within the social fabric (Garcia 1981; McIntosh, Tainter, and McIntosh 2000a; Tainter 1988, 1995). These internal weaknesses are amplified in the presence of external stress, eventually leading to social collapse. The same underlying factors would have responded in a similar way to other external pressures including disease and warfare as well as environmental change.

An example of this alternative way to view the relationship between environmental change and society comes from McGovern's (1994) work on the medieval Norse populations that inhabited Greenland. These people first colonized this region in the tenth century AD during a phase commonly known as the Medieval Climatic Optimum, which preceded the much-studied cold phase of the Little Ice Age. At this time conditions were well suited to a dairy-herding lifestyle, and the community prospered for several hundred years. Beginning in the late thirteenth century, the cold climatic conditions of the Little Ice Age set in and led to a continual decrease in pastoral productivity and the worsening of economic conditions for this Norse population. At the same time, there were Inuit communities living on the island who were adapted to a seal-hunting economy. With the increasingly colder conditions the sole change in the Inuit economy was the variation in ratios of the exploitation of different seal species. The Norse population, however, suffered major economic strain, which manifested itself in several forms of social stress. Instead of becoming more adaptable by exploiting new technologies or resources as the Inuit did, the Norse ruling elite tightened their control on the population by stressing cultural conservatism. There were more witch trials than there had ever been before, and vast resources were spent building cathedrals and other religious manifestations. By the end of the fifteenth century AD the Norse population had completely vanished from Greenland.

The success or failure of a given community's ability to weather severe environmental change is related to such internal factors as social organization, technology, and the perception of environmental change and its causes. Environmental perception very much influences the understanding of measures to be taken to overcome detrimental environmental shifts. For example, if an episode of severe drought is attributed to an incursion of a foreign population in a given region, then the solution to this drought would be perceived as related to the expulsion of this outside element. Such a situation occurred in fifteenth-century Spain, where a series of drought years led to violent riots directed at the minority Moslem population (Mackay 1981). These people were perceived as displeasing the Almighty, who manifested this displeasure by inducing rainfall failure. Similarly, if the cause of environmental change is attributed to the sins of the community, as in the case of the Greenland Norse population, then relief from a detrimental climatic shift might come from intensified religious activity in the form of increased investment in religious institutions and the strict enforcement of traditional rituals. This is nicely illustrated in biblical texts, in which prophets admonish the population for their sins and threaten divine punishments entailing the cessation of rainfall and drying up of the agricultural land (Carroll 1986).

Thus, the culturally perceived formula for relief from environmental degradation would in some cases have less to do with technological change and enhancement than with internal social factors. This is not to say, however, that technological change can never be a solution employed by preindustrial societies. In the same sense cosmological solutions are not exclusively applied to solve agricultural and economic stress. As Escobar remarks, "It is necessary to strive for a more balanced position that acknowledges both the constructedness of nature in human contexts . . . and nature in the realist sense, that is, the existence of an independent order of nature" (1999:3). Human responses to environmental change are rarely exclusive to one or the other of these conceptual frameworks; however, they are weighted differently in different societies.

We can contrast the manner in which preindustrial societies, such as that of the Iron Age peoples of the Southern Levant, dealt with environmental fluctuation with the manner in which twentieth-century Western societies approach the problem of global warming and severe weather events associated with El Niño. The modern Western focus is on developing new agricultural technologies and reducing pollution-causing greenhouse gasses (Nilsson and Pitt 1994; Turner and O'Connell 2001). According to Iron Age Israelite texts, the solution for solving the drought problem focused on controlling individual and community behavior displeasing to the supernatural being. This is not to say, however, that these two very different societies viewed the environment as an exclusively cultural construct or as an extracultural entity or essence (Escobar 1999). In addition to their belief systems the ancient Israelites of the Iron Age had very sophisticated water storage technologies and terraced agricultural systems (Borowski 1987), whereas communities in our own technocratic Western society take to their places of worship during environmental disasters such as floods and droughts (see, for example, Egan 2002). But it is the cultural system itself that lends greater importance to one perceptual framework over another. Therefore, in environmental archaeological studies it is important to keep in mind that the perceptions and responses of ancient societies to environmental stress might not be similar to the way in which modern Western societies approach these problems. The successes or failures of these different approaches are not always related to one construct or the other.

There are also cases in which societies are robust enough to survive either intensely disruptive social changes or severe episodes of drought, but the convergence of these two unfortunate situations leads to social and political upheaval and change. Such a case is nicely illustrated by the period of the Pueblo Revolt of 1680 in the American Southwest. From the time of initial contact in 1539, the Spanish viewed the Pueblos as an exploitable resource for supplies and labor. This asymmetrical interaction continued to escalate at a faster pace with the permanent settlement by Spanish missionaries and colonists who arrived from central Mexico in 1598 (Knaut 1995:18). The Spanish subsisted primarily upon provisions supplied from Pueblo storehouses.

Previous to the Spanish incursion the Pueblos managed to survive in a drought-plagued semiarid environment by instituting a system of buffers that allowed them to overcome periods of severe drought. These buffers included floodwater farming of maize, sophisticated storage technology, trade between pueblos and nomadic peoples from the Great Plains, social and kinship ties with other Pueblo groups, and the ability to resettle into aggregated towns close to areas of highest productivity during extended periods of drought (Dean 2000; Jorde 1977). In addition to these more material responses to a drought-plagued climate, there were also social and cosmological adaptations. One of the major phenomena to emerge from the period of great environmental degradation of 1270–1450 AD in the Southwest was the Kachina cult (Adams 1991:120–121; Dean 2000). Kachinas were perceived as the beings who brought rain, and these entities were integral to major rain-inducing ceremonies (Adams 1991:9).

The arrival of the Spanish brought new demands upon the Pueblo economy and also new world views in the form of seventeenth-century Spanish Catholicism. The draining of the Pueblo storehouses was further exacerbated in times of drought that occurred periodically throughout the seventeenth century (deMenocal 2001). The ultimate result was a complete erosion of the Pueblo buffering system. Many Pueblos were suddenly dependent upon the Spanish, who now controlled the supplies of maize. This may have been one major incentive for their initial conversion to Christianity (Knaut 1995:64; Silverberg 1970: 48). The Spanish suffered as well from the droughts since they were not producing anything close to substantial quantities of supplies and depended almost entirely on the Pueblo production systems.

Severe drought conditions set in around 1660 and lasted over the course of almost twenty years (Knaut 1995:157; deMenocal 2001; Silverberg 1970:88). At the very time when Pueblo groups were experiencing the most need to return to a belief system that they perceived would ease the suffering from the droughts, the Spanish tightened their grip on the community. In 1661 and again in 1675 Governor de Treviño attempted to enforce conformity to a Christian way of life by widespread burning of traditional Pueblo ceremonial objects, the destruction of kivas, and the arrest of forty-seven Pueblo spiritual leaders. These actions initiated the Pueblo Revolt in 1680, which temporarily ended eighty years of Spanish rule in New Mexico (Folsom 1973; Knaut 1989; Reff 1995).

In studying the history of this revolt we can attempt to understand the role played by environmental degradation. The extended period of drought was not a sole factor in inducing the native population to rise up against the Spanish. There had been severe droughts throughout the first eighty years of Spanish domination and a number after the period of the 1680 Pueblo Revolt. One of these, around 1740, was just as intense and lasted significantly longer than that of the 1670s (deMenocal 2001). However, after 1680 there were no more revolts against Spanish rule in the Southwest. This emphasizes the point that the breakdown in Spanish colonial rule in 1680 was caused both by the drought and by a number of social factors resulting from decision making on the part of the Spanish and the Pueblos. If the Spanish had taken a different tactic and directed public works toward improved irrigation systems, a fair system of food redistribution, and an attitude of spiritual tolerance, they might well have improved their relations with the Pueblos and circumvented the culmination of hostilities into a revolt.

SUCCESS AND FAILURE IN THE WAKE OF CLIMATIC CHANGE

Archaeologically there has been little attention paid to those societies that have successfully survived severe environmental stress or to the reasons behind this achievement. Much more notice is given to apparent cases of environmental pressure and collapse, which has the appeal of catastrophism and climatic sensationalism especially in this modern era of concern over global warming and El Niño–related flood events (Weiss and Bradley 2001). Successful adaptations to environmental stress are much less visible archaeologically unless one is specifically looking for such situations. The question of why some societies deal successfully with environmental stress while others are unsuccessful is of far more anthropological interest than simply arguing a case for climatic stimulus leading to a collapse response.

Conversely, periods of climatic amelioration are not necessarily guaranteed times of wealth, cultural florescence, and population expansion. With close examination it is possible to note underlying cultural influences that lead to these increases in social and technological complexity. Favorable climatic conditions are important in that they provide opportunity, but they are not prime movers that dictate the outcome (Issar and Zohar 2004). An example that illustrates this is a case of agricultural intensification among central Asian nomads during the Iron Age Scythian period in southeastern Kazakhstan (Rosen, Chang, and Grigoriev 2000). The early part of the Iron Age beginning in the tenth century BC was characterized by cool climatic conditions with variable humidity (van Geel et al. 2004). Initial settlement of southeastern Kazakhstan by Saka Scythian groups took place during this cool episode. These people were nomadic herders of sheep, goats, and horses who arrived in the area presumably with only rudimentary agricultural technologies.

Sometime around the third century BC climatic amelioration and considerable warming began to take affect (Khotinskiy 1984; Rosen, Chang, and Grigoriev 2000). Phytolith evidence from agro-pastoral Saka-Wusun sites show an increasing trend toward more intensive agriculture from the fifth through the third century BC, apparently beginning before this amelioration took effect. This intensification is most likely to have been related to historical factors grounded in the consolidation of nomadic confederacies to the east (as a result of hostile contact with the Han Chinese empire), the pushing of Wusun tribes into the Saka territory, and the formation of professional warrior classes. Climatic amelioration provided the opportunity for more intensive cultivation but was not the solitary cause. It is more likely that the warrior and elite social classes created a need for a more stable food supply than would have been necessary for migratory semiegalitarian pastoral groups, and the warming climatic conditions enhanced the success of these endeavors. Thus the adoption of more intensive agricultural strategies was more than a function of ameliorating climate. The warmer climatic conditions provided a "pull" factor, but the "push" factor derived from the social need to support an increasingly more complex society among the seminomadic peoples of this region.

ORGANIZATION OF THIS VOLUME

In this volume I examine the role climate has played in Near Eastern society from the perspectives outlined above. To do this it is first necessary to review the procedures used in determining the evidence for climatic change and then to summarize the evidence itself to reconstruct the Late Pleistocene and Holocene climatic sequences as far as they can be estimated from the current state of our knowledge. I have summarized some of the defining evidence for climate change at both a detailed and a broad-brush level. Readers who wish to acquire only a general background to the sequence of climatic events in the eastern Mediterranean Levant can restrict their reading to the general summaries by period located at the end of chapters 4 (the Pleistocene) and 5 (the Holocene). Those who want a more detailed summary and interpretation can read the particulars and perhaps use these as jumping-off points for further study. Radiocarbon dates throughout the volume are reported as calibrated dates where possible, as either years before present (BP) or calendar years (BC or AD). Where dates are uncalibrated, they are reported with a lowercase "bp."

After the chapters summarizing Pleistocene and Holocene climatic change I examine three specific problems of human-climate relations in more depth. The first deals with hunter-gatherers at the end of the Pleistocene, and the relationship between climatic change and the origins of plant domestication. The second case study pertains to the onset of severe sustained drought conditions and their impact on both resilient and rigid complex societies at the end of the Early Bronze Age in the Levant. Finally, I examine the effects of both benevolent and detrimental climate on the Roman and Byzantine empires that controlled vast areas of the Near East. The intention is to present case studies that might serve as models of human action for similar societies worldwide given similar aspects of social organization, political structure, technologies, and environmental setting.

Tools for Understanding Paleoenvironments in the Southern Levant: Principles of Climatic Reconstruction in the Near East

Researchers measure past climates in a number of ways. Modern climatic change can be determined with the help of instruments that obtain direct records of air and sea temperatures, wind direction, and rainfall amounts and frequencies. Such records, however, have been available to us for only approximately four hundred years beginning with the invention of the barometer by Torricelli and the thermometer by Galileo in the first half of the seventeenth century AD (Lamb 1995). In the Near East, modern systematic rainfall and temperature records do not go back much further than the past century. In place of direct climatic data, researchers use what is known as *proxy* data. These are records of natural phenomena that have been directly impacted by climatic change, and therefore they themselves provide a secondary record. These data sets include *historical accounts*, such as records of fluctuating wheat harvests; *paleoecological data*, including fossil pollen, faunal records, and foraminifera in marine sediments; and *geological data*, such as sedimentation rates in lakes and valleys, stream terrace formation, and isotopic analysis of oxygen and carbon in marine sediments (see table 2.1).

A number of researchers have also used archaeological evidence for fluctuations in settlement and abandonments as a form of climatic proxy (see Bruins 1990; Issar and Zohar 2004). This can sometimes be a valid approximation of environmental fluctuations; however, it must be used with extreme caution and usually is more reliable when restricted to hunter-gatherer populations, which were generally mobile and resilient societies. It is very much the thesis

Table 2.1.   Sources of Proxy Data for Paleoclimate Reconstructions (After Holliday 2001:21, Table 1.2)

Glaciological (ice cores)
    Oxygen and hydrogen isotopes, major ions
    Gas content in air bubbles
    Trace elements and microparticle concentrations (e.g., dust)
    Physical characteristics
Geological
    Marine (ocean sediment cores)
    Microfossils
    Oxygen isotopes (of foraminifera)
    Sediment mineralogy and geochemistry
    Eolian dust and pollen
    Ice-rafted debris
    Clay mineralogy
Terrestrial
    Glacial landforms and deposits
    Periglacial and other mass-wasting landforms and deposits
    Eolian landforms and deposits
    Fluvial landforms and deposits
    Lacustrine landforms and deposits
    Cave deposits
    Mire and bog deposits
    Soils and other weathering characteristics
Biological
    Tree rings
    Pollen
    Phytoliths
    Corals
    Plant macrofossils
    Invertebrate fossils
    Vertebrate fossils
Historical
    Weather records
    Weather-dependent phenomena
    Phenological records

of this volume that in addition to climatic change more complex societies are controlled by a number of other social and technological factors that work in concert to influence settlement and abandonment. Therefore, the expansion and contraction of these settlements and urban centers cannot provide a clear or simple proxy for climatic fluctuations.

It is extremely important for archaeologists, historians, and students to understand the strengths and weaknesses of the battery of proxy data sources. These data are often subject to subjective interpretation and fudging in the same way as seemingly less *objective* data produced by social science techniques. Many archaeologists unquestioningly accept interpretations of these data without realizing which are the subjective and which are the objective elements within a summary of pollen and geomorphological results. Therefore, archaeologists should take a critical look at all proxy climatic data that they cite when using a "climatic change" explanation for cultural developments. With this in mind, I briefly review some of the more common methods used for climatic interpretation in the Late Pleistocene and Holocene in the Near East. Many methods are available to researchers of climatic change, but I review only the techniques that are most frequently used in the region. A more comprehensive discussion of methods for Quaternary climatic reconstruction can be found in Bradley (1999), Lowe and Walker (1997), and Williams et al. (1998).

HISTORICAL RECORDS

Historical records can provide indirect information about changing climatic conditions. Ingram, Underhill, and Wigley (1978) group historical data into a number of categories including (a) ancient inscriptions, (b) annals and chronicles, (c) government records, (d) private estate records, (e) maritime and commercial records, (f) personal papers, such as diaries or correspondence, and (g) scientific or protoscientific writings, such as (noninstrumental) weather journals (cited in Bradley 1999:441).

Some of the historical records used for climatic reconstruction in the ancient Near East include accountings of crop yields and types of plants cultivated (Jacobsen and Adams 1958), tax records, and Roman, Talmudic, and Byzantine accounts (Mayerson 1967; Sperber 1978). These contain general observations of unique weather events such as droughts, floods, and severe frosts. Unfortunately, these types of records suffer from a lack of reliability. Not all textual accounts are as accurate as those of the present, due to the absence of precise instruments, lack of regularity in written observations, and the inaccuracies of evaluating adjectives such as the "worst" winter, "wettest" year, or "driest" spring. Also, observers might not always have been dependable in their accounts of the magnitude of events and the timing or periodicity of their occurrences. Some of the chroniclers have moralistic motivations, and the experiences recounted are elaborated to emphasize such philosophies as the manner in which the fertility of the land is destroyed by droughts and floods when the inhabitants engage in immoral behavior (Carroll 1986).

POLLEN ANALYSES

One of the foremost paleoecological methods is the analyses of pollen sequences from lake and bog cores. Palynology is a technique that has been used by geologists for stratigraphic correlation and climatic reconstruction since the nineteenth century (Flint 1971:25). Pollen grains are microscopic particles ranging in size from 10 to 100 μm with most falling within the 25–35 μm size range (see figure 2.1). They are produced by the reproductive parts of seed-producing plants (angiosperms and gymnosperms) and distributed in large numbers by either wind, insects, or other more incidental carriers such as water and animals. The outer layer of pollen grains is composed of a durable organic substance known as *sporopollenin*, which preserves well in damp anaerobic lake sediments or hyperarid conditions for hundreds of thousands of years. The forms of individual pollen grains are usually distinctive at the genus level of taxonomy and more rarely at the species level. They are identified using criteria based on size, shape, surface morphology, and number of apertures. Pollen is produced in immense numbers by trees, shrubs, and grasses. Wind-born pollen can travel for hundreds of miles before being incorporated within the surface sediments, lake, or bog deposits.

Pollen is extracted from lakes, marshes, and peat deposits by a process of coring with specialized equipment. In ideal conditions the coring rigs can penetrate tens of meters of fine-grained sediments. Once extracted the sediment cores are sealed and kept cool to protect the pollen from destruction through oxidation. In the laboratory the cores are sampled at varying intervals, and material is collected for radiocarbon dating. The number of points dated in the core is restricted to the presence of dateable carbon in the form of charred wood, reeds, or mollusk shells, and the funding available for such dates. It is desirable to have as much of the core dated as possible to have a better timescale for vegetation changes and a good understanding of the changing rates of sedimentation in the lake or marsh basin.

The pollen samples themselves are processed through a series of refining procedures that are designed to remove the matrix of mineral matter and extraneous organic materials. The diagenetic carbonates and calcareous detritus are removed with hydrochloric acid. Silt and clay minerals are eliminated using hydrofluoric acid to dissolve the silica, or the lighter pollen and other organic material is floated in a heavy-density liquid and collected. Extraneous organic materials are chemically oxidized with hydrogen peroxide (H_2O_2). The remaining pollen and spores are stained and mounted on microscope slides.

The various pollen forms are counted up to a preestablished number for statistical accuracy. This is usually somewhere between 300 and 1000 individ-

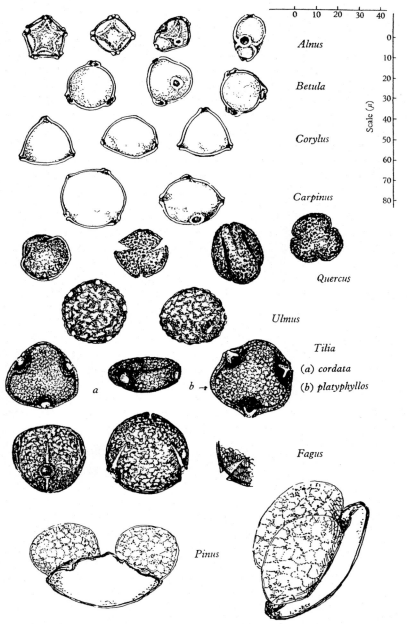

FIGURE 2.1
Illustration of Various Pollen Grains (After Godwin 1956).

uals. The total count is referred to by palynologists as the *pollen sum*. These counts can be expressed either as relative percentages of a total or as absolute numbers of pollen per unit of sediment. The disadvantage of using relative percentages is the *interdependence* of the results for the different taxa. For example, if oak, pine, and pistachio types are the only ones counted, then their total percentages would add up to 100%. If the quantities of pine pollen fluctuated, then the percentage of pine would change, thereby altering the percentages of the other components of the pollen assemblage whether the quantities of these other forms changed or not.

Absolute counts give a more accurate, independent accounting of the amounts of each taxon of pollen in a given unit of sediment. These counts are usually done by spiking a sediment subsample of known volume (e.g., 1 cm³) with a known number of an exotic pollen or spore type (sometimes *Lycopodium* spores in North America, or *Eucalyptus* pollen in Europe), and calculating the pollen numbers in relation to the number counted of this foreign element. Thus if one counts only 20% of the known total of exotic pollen, then all the fossil pollen types counted are also only 20% of the total pollen in that subsample of a cm³ of sediment. This is a much more precise method for determining pollen presence in a core. The reader can find more detailed information about pollen types, laboratory procedures, and pollen counts in Faegri and Iversen (1989) and Moore, Webb, and Collinson (1991).

Pollen diagrams illustrate pollen numbers and/or percentages in samples from stratified sediments. They are presented as curves for each pollen category displayed in columns, and they record the progression of vegetation changes through time (see figure 2.2). Some pollen forms indicate wetter or drier conditions, and others point to cooler or warmer environments. The major phases of pollen change are divided up into *pollen zones*. These zones are used in interpreting changes of vegetation that may have been influenced by climatic changes or human land use. Another tool used by pollen analysts is the curve for a*rboreal* versus *nonarboreal* pollen. In the Eastern Mediterranean Basin, and the Near East in general, a rise in the arboreal pollen curve is construed as a proxy for moister climatic conditions. Conversely, a rise in nonarboreal pollen, primarily dry-steppic shrubs such as Chenopodiaceae and *Artemisia*, indicate the onset of a dry climate.

There are several pitfalls in the interpretation of pollen data of which archaeologists should be aware. One of these is that the radiocarbon dates published with pollen sequences are often uncalibrated. Therefore, archaeologists

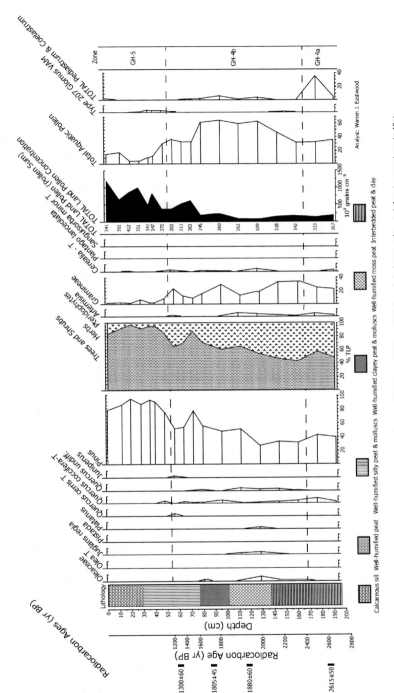

FIGURE 2.2

Example of a Pollen Diagram. Note Sediment Units, Pollen Zones, Percentages Indicated on Lower Line, and Arboreal Versus Nonarboreal Curves (Redrawn from Eastwood et al. 1999).

Summary percentage pollen diagram for core GHC from Gölhisar Gölü for selected taxa only.   '+' denotes rare taxa (< 1%).

Analyst: Warren J. Eastwood

need to calibrate these into calendar years before being able to compare the data with archaeological time periods. Another difficulty in interpreting radiocarbon dates from lake cores occurs in regions with an underlying geology of carbonate rocks or recent volcanic activity. The eroded particles from these rocks and their associated sediments are rich in "old carbon." The old carbon enters the lake and alters the (carbonate) chemistry of the lake water. This in turn affects subaquatic photosynthesis of water plants and carbonate precipitation by freshwater organisms such as mollusks. The phenomenon is responsible for creating a "hard water error" on the radiocarbon dates that can add up to 1200 years to the radiocarbon ages from mollusk shells and the remains of water plants (Lowe and Walker 1997:245–246). This trend is discussed in more detail in chapter 3, as it pertains to pollen diagrams in the Southern and Northern Levant.

In interpreting pollen diagrams, archaeologists must also take into consideration the lag time embodied in vegetation changes. With climatic amelioration, the movement of a forest into an area can take several hundred years or more, depending upon where the refuge of the forest was, the distance from the region where the pollen core was collected, and the intervening environmental zones through which the forest would have had to "travel." Thus the climatic changes implied by the pollen indicators of forest colonization could have occurred long before they are reflected in the pollen rain.

Another problem faced by palynologists is the variation in the amounts of pollen distributed by different modern plant species. Pine, for example, disperses far more pollen than oak over long distances, and pistachio—a prime indicator of the warm wet conditions that ushered in the Holocene—is a notoriously low producer of pollen. This factor must be accounted for in the interpretation of pollen diagrams. It is easy to make the mistake of falsely reconstructing proportions of vegetation in the vicinity of a pollen core. This false reconstruction can in turn lead to faulty determinations of climatic sequences. Another important element of confusion is the factor of *human impact* on vegetation. The impact of the human element increases throughout the Holocene with the greatest extent of influence appearing in the Late Holocene. This effect on vegetation is so significant that there are very few pollen diagrams from the last 2000 years that are of use for direct climatic reconstruction. Even before the major environmental changes that came about with increasing agricultural intensification, humans have impacted vegetation by the use of fire in forest clearance for low-intensity farming, hunting, foraging, and pastoralism. Therefore, it is rare to be able to assume a case of pristine climax vegetation from any area in the Late Pleistocene and Holocene (Roberts 1998:35–36; Lowe and Walker 1997:175).

ISOTOPE ANALYSES

The use of isotope analyses as proxy records for climatic change has rapidly become one of the most powerful techniques available to paleoenvironmental researchers. Isotopes are forms of a chemical element that have the same atomic number (sharing the same number of protons), but different numbers of neutrons in the nucleus. The most well-known isotope to archaeologists is carbon-14, an unstable isotope used for dating organic materials found in archaeological sites. One of the most important isotopes for environmental studies is oxygen-18, which is a rarer isotope than the more common oxygen-16. The ratio of ^{18}O to ^{16}O ($\delta\,^{18}O$) is an indicator of temperature change from sea and ice cores (Bradley 1999) and in some depositional environments such as lakes can be an indicator of relative humidity due to the increase in $\delta^{18}O$ levels with evaporation. In cave speleothems, $\delta^{18}O$ can be calibrated to indicate fluctuations in the amount of precipitation. Oxygen isotopes have been used in analysis of the climatic record from the Greenland ice pack, sea and lake sediments, speleothems from caves, and snail shells extracted from ancient alluvial sediments. The $\delta^{18}O$ values from ocean and sea floors come from biogenic $CaCO_3$ formed in the shells of fossil marine planktonic foraminifera.

The lighter oxygen (^{16}O) evaporates from ocean water more readily than the heavier ^{18}O. This effect is enhanced under colder climatic conditions. ^{16}O is also the first to precipitate from the atmosphere as rain or snowfall. During colder climatic episodes, ^{16}O evaporates from the ocean water and is locked into the ice of glaciers and continental ice sheets, thus depleting the sea water of the lighter oxygen and resulting in higher $\delta^{18}O$ values. In warmer phases the ^{16}O is released with the ice melt thus producing lower $\delta^{18}O$ values in sea cores. The converse applies to ice cores with the lighter ^{16}O accumulating in ice sheets during cold periods and heavier in warm phases, resulting in lower $\delta^{18}O$ levels during glacial episodes and higher values during interglacial periods (see figure 2.3).

In terrestrial sources such as lake sediments, land snails, and speleothems, lower levels of $\delta^{18}O$ indicate cooler environmental conditions due to the input of isotopically lighter rain water. The $\delta^{18}O$ values from cave speleothems are commonly used as temperature indicators (see, for example, Lauritzen 1995;

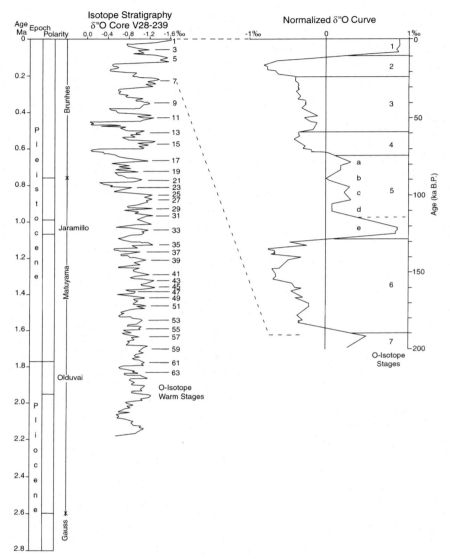

FIGURE 2.3
Oxygen Isotope Stages from Sea Cores (After Holliday 2001:8, Figure 1.2).

Winograd et al. 1997). However, isotope records from speleothems are more difficult to interpret than those from sediments from other terrestrial settings due
to the unique conditions under which they form within cave environments
(Bradley 1999:329–334). Speleothems form from calcite deposits that precipitate out of water dripping from cave walls and ceilings. Since air and water
movement in a cave is relatively slow, the temperature of the air and the bedrock
should theoretically reach an equilibrium that approximates the mean annual
surface temperature. The fractionation of oxygen isotopes from the calcite
speleothems is dependent upon temperature and therefore should provide a
record of surface paleotemperatures at the time of speleothem deposition. Because of possible variations in local cave environments, before undertaking such
analyses it is necessary to measure consistency of $\delta^{18}O$ along a single growth
layer as well as the relationship between the carbon and oxygen isotopes. If there
is a correlation between the carbon and oxygen isotopes, then the fluctuations
were controlled by kinetic factors indicating that the speleothem was deposited
in a situation of nonequilibrium, and therefore the results would not be useful
for determining paleotemperatures of the surface. This can be tested by analyz-
ing the $^{18}O/^{16}O$ ratios in modern groundwater and in the calcite that precipi-
tates from it (Bradley 1999:330). The interpretation of the $\delta^{18}O$ values can also
be complex, but it is generally accepted for most cave environments that there
is an *increase* in $\delta^{18}O$ with a *decrease* in cave temperature.

Bar-Matthews and her team have conducted extensive work on $\delta^{18}O$ sequences from Levantine caves that have yielded critical paleoenvironmental
data for the Late Pleistocene and Holocene (Bar-Matthews, Ayalon, and Kaufman 1997, 1998; Bar-Matthews et al. 1999). In this work they found a stronger
correlation to fluctuations in rainfall amounts than to temperature changes,
and therefore their results from $\delta^{18}O$ determinations on speleothems at the
Soreq Cave are based on calibrations of the $\delta^{18}O$ signal with rainfall fluctuations. These values from the Soreq Cave correlate with isotope data from the
Eastern Mediterranean, where Schilman et al. (2002) demonstrated that the
variation in rainfall amount has a much greater affect on $\delta^{18}O$ values than sea
surface temperature, and thus they conclude that their isotope data from both
sea core and speleothem sources yield more significant information about
changing precipitation throughout the Pleistocene and Holocene than about
temperature fluctuations (Schilman et al. 2002).

Carbon isotopes represent another method for determining environmental
conditions. In the Southern Levant, the ratios of $^{12}C/^{13}C$ ($\delta^{13}C$) have been

used in analyses of snail shells to provide environmental information from the types of plants consumed by the snails. There are two carbon pathways of photosynthesis used by plants, designated C_3 and C_4. Plants that utilize the C_3 pathway generally grow in more mesic conditions than C_4 plants. The $^{12}C/^{13}C$ ratios indicate which types of plants the snails were consuming. This is a powerful indicator of environmental changes, particularly in desert margins, where vegetation is very sensitive to subtle shifts in rainfall amounts (Goodfriend 1999).

GEOMORPHOLOGY

Geomorphology is the study of modern and ancient landforms. The landscape—including hill slopes, valleys, stream deposits, and shorelines—is very sensitive to natural as well as human-induced environmental changes. These environmental shifts leave a record behind in landforms and sediment deposits, which can then be reconstructed by earth scientists and interpreted for information about past climatic regimes and human land-use practices. Landscape features are composed of a variety of different sediment types. The sediment particles themselves progress through life cycles that include phases of weathering, transportation, deposition, and diagenesis. All of these phases leave their traces at each stage. Observation of the characteristics imprinted on the sediment deposits can ultimately furnish considerable information about past climate, hydrology, vegetation cover, and human impact.

Important characteristics include sediment color, grain size, sorting, texture, and composition. *Color* can indicate weathering and oxidation in reddish sediments, waterlogging and anaerobic conditions in bluish or greenish sediments, or marshy microenvironments in organic-rich black deposits. *Grain size* is the size of the sediment particles, described according to the Wentworth Scale as clay (< 0.002 mm), silt (0.063–0.002 mm), sand (1.00–0.063 mm), and gravel (> 1.00 mm) (Krumbein and Sloss 1963:96). The size of sediment particles is an indicator of the energy of the depositional environment, with quiet lake environments depositing fine silts and clays, as opposed to more powerful streams depositing sands and gravels (see figure 2.4).

Sediment sorting is also a good indicator of depositional environment. Sorting refers to the relative percentages of gravel, sand, silt, and clay in sediment. The more homogeneous the deposit the better its sorting. Good sorting occurs in well-sustained transport media as in windblown silt or sand, and poor sorting is typical of highly variable flows as in alluvial sediments deposited during flash flooding. *Sediment texture* is the blend of particle sizes that gives the deposit substance. Textures are described as "sandy silt," "clay loam," "boulder clay," and so on. The *sediment composition* also provides important clues to the origin of the deposit. Sediment sources may vary between stream catchments. For example, a localized basalt deposit from a specific time period may exist in the upper reaches of a given stream network. If stream deposits in this catchment contain no basalt gravels, then the alluvium

FIGURE 2.4
Sediment Deposits from a Roman Period Stream near Gordion, Turkey (Photo Credit A. Rosen).

may predate the basalt flow. Anthropogenic materials within a deposit may also indicate the impact of local land use.

Soil types also contain much information about past climates. Mediterranean *terra rossas* are distinctive relicts of Pleistocene forest soils that developed under moister conditions than are now extant in the Near East. Signs of soil development are also clear indications of stable land surfaces. These can occur in sediment sections on semiactive floodplains and signify interludes of landscape stability with no significant alluviation or erosional episodes.

On a larger scale, there is much climatic information to be obtained from the study of stream terrace systems as well as from the sediments they contain. These relicts of abandoned floodplains can form a series of stepped land surfaces defining the transverse profile of a stream valley. Stream terraces record higher stream flows that dominated the flow regime in the past and the location of the zone of sediment deposition. They preserve a record of various facies of stream deposits including sediments from channels, bars, levees, and old floodplains (see figure 2.5).

All of the above-mentioned types of deposits are highly sensitive to changes in rainfall, temperature, vegetation, and topography. As such, they

FIGURE 2.5
Large Late Pleistocene Stream Terrace from Kadesh Barnea, Sinai (Photo Credit A. Rosen).

can provide much information about the history of climatic fluctuations in a given locality.

CONCLUSION

The climatic proxies discussed above are far from a comprehensive list of the battery of techniques available for climatic reconstruction. However, these are the techniques that have been most frequently employed in the reconstruction of climatic regimes in the Near East. As such, it is important for archaeologists, historians, and students of human societies in the Near East to understand how climatic determinations are made from the available evidence. Further reading on these topics can be pursued in some of the references cited above. In this way, social scientists can more readily evaluate the contributions of each of these proxy methods.

3

Land and History: Introduction to the Modern Landscape and Historical Framework

GEOGRAPHIC SETTING

The Southern Levant as defined for the purposes of this study comprises much of the territory of Israel, Palestine, the Egyptian Sinai, and Jordan. The latter is often referred to as Transjordan, meaning the lands east of the Jordan River. The geography of the region has much to do with its geopolitical history. The Southern Levant has always formed a land bridge between Africa, Europe, and Asia, with the Mediterranean Sea to the west and the Arabian Desert to the east. In spite of the relatively small size of the territory, it encompasses a mosaic of different environmental zones that provide opportunities for the production and exploitation of a variety of resources.

Some of the main features of this landscape include the hyperarid Sinai Desert, the semiarid northern Negev Desert, the relatively broad and flat coastal plain, and the longitudinal backbone of low mountains and *Shephela* foothills stretching between the coastal plain to the west and the Jordan Rift Valley to the east. The northern portion of this territory encompasses the forested hilly zone of the Galilee. The Rift Valley is divided into the Jordan River Valley to the north, beginning just south of Lake Kinneret (Sea of Galilee) and extending south to the tip of the Dead Sea. From the Dead Sea to the Red Sea is the Arava Valley.

Climatically, most of this region falls into the Mediterranean zone. This is characterized by cool wet winters and warm dry summers (Orni and Efrat 1980). The coldest month is January with low temperatures averaging from about 9°C in the hills to about 13°C in the Jordan Valley, and the warmest month is August with temperatures averaging about 24°C in the hills and about 34°C in the Jordan Valley. Extreme heat can occur at intervals during the summer months with temperatures hovering around 40°C. These hot dry *sharav* conditions can be beneficial for agriculture because they accelerate the ripening of grains and fruit and have a negative effect on weeds and pests. With irrigation, farmers are able to stagger harvest dates and therefore ensure that crop supplies are available for longer periods of time during the year (Orni and Efrat 1980). Annual rainfall averages range from lows of about 15 mm in Eilat in the south of the country through about 200–300 mm in the northern Negev, about 550 mm in the central hill region, and up to about 800 mm in the northern Galilee.

The soils vary from silty loess soils in the northern Negev, to now-eroded terra rossa soils in the hills, and heavy black vertisols in the inland valleys and coastal plain. The streams for the most part are seasonally active wadis, which are deeply entrenched and dry throughout the summer months.

Vegetation is usually divided into four major vegetation zones including (a) Mediterranean, (b) Irano-Turanian, (c) Saharo-Arabian, and (d) coastal dune vegetation with additional small patches of tropical vegetation (the Sudano-Deccanian zone in parts of the Jordan Valley and Arava). The Mediterranean zone covers most of central and northern Israel and is divided into the coastal plain/inland valleys and hills. The hills were traditionally covered by oak (mostly *Quercus calliprinos*)/pistachio (*Pistacia palaestina*) forests and high maquis, and include some pine (*Pinus halepensis*) and carob (*Ceratonia siliqua*) as well. Tabor oak (*Quercus ithaburensis*) grew in the central and northern sections of the coastal plain. The Irano-Turanian vegetation zone is located along the semiarid fringes of the central region, including the northern Negev on loess or calcareous soils. This area is typified by wormwood (*Artemisia herba albae*) and some degraded steppe grass species. The Saharo-Arabian zone includes much of the Negev, the Judean Desert, the Dead Sea region, the Arava Valley, Sinai, and the southern and eastern portions of Transjordan. The vegetation is sparse and concentrated only in wadi beds. It includes such shrubs as the bean caper (*Zygophyllum dumosi*) and the broom plant (*Retama roetam*) (Orni and Efrat 1980).

ARCHAEOLOGICAL BACKGROUND

This volume does not attempt to examine the entire course of human-environmental relations in the Southern Levant throughout the Late Pleistocene

and Holocene. To do so would be to take a superficial view and gloss over some
of the more important issues that might be of benefit to other similar studies
of human-climate relations in varying localities throughout the world. As
stated in the introduction, I have therefore selected three critical time periods
and levels of human social organization including (a) the Terminal Pleistocene
Natufian hunter-gatherers, (b) the Middle Holocene early complex societies of
the Early Bronze III period, and (c) the Late Holocene Roman-Byzantine Empires. To put these periods in their archaeological context it is necessary to provide a general overview of the archaeological periods in the Southern Levant.
The general chronology of the periods is listed in table 3.1.

The Terminal Pleistocene is marked by increasingly more complex hunter-
gatherer societies such as the Geometric Kebaran and Natufian, which inhabited the Mediterranean forest zones in the Epipaleolithic period (from ca.
16,000 to 11,600 cal. BP). The people of the *Geometric Kebaran* complex lived
in the Mediterranean zone from circa 16,000 to 15,000/14,600 cal. BP and left
behind the remains of small, nomadic sites with insubstantial round structures. The society consisted of small family-based mobile groups, and it is rare
to find burials in this period. In addition to their chipped-stone industries,
their tool kit included ground-stone grinding equipment, shell beads, and rare
bone implements.

In the Negev and Sinai, there are clusters of Geometric Kebaran sites of various sizes with larger sites at higher elevations. There is apparently little evidence of seasonality influencing site size or location; however, all the sites are
close to major water sources and according to Goring-Morris and Belfer-Cohen (1998) are of Mediterranean derivation. These people hunted mostly
gazelle, but also hartebeest, aurochs, and onager.

The Geometric Kebaran complex was replaced in the Negev and Sinai by
the Mushabian and later the Ramonian complexes, but in the more
northerly Mediterranean zone, they continue until Natufian times beginning at circa 15,000/14,600 cal. BP (Bar-Yosef 2002a). Climatically, the Geometric Kebaran period is an episode of general warming and increasing
humidity after the height of the cold dry Late Glacial Maximum (LGM) of
circa 22,000 cal. BP.

The *Natufian* culture is characterized by a number of different regional entities and spans a period of over 3000 years. Archaeologists have divided it up
into phases of Early, Late, and Final Natufian (Bar-Yosef 2002a; Goring-Morris 1998, Valla 1998a). Natufians are the subject of much study since they are

Table 3.1.   Chronology for the Southern Levant and Neighboring
Regions (After Bar-Yosef 1998; Dever 2003; Kuijt and Goring-
Morris 2002; T. Levy 1998a).

Late Epipaleolithic
   Early Natufian   14,500–13,000 cal. BP
   Late/Final Natufian   13,000–11,600 cal. BP

Neolithic
   Pre-Pottery Neolithic A (PPNA)   11,700–10,500 cal. BP
   Pre-Pottery Neolithic B
     Early PPNB   10,500–10,100 cal. BP
     Middle PPNB   10,100–9250 cal. BP
     Late PPNB   9250–8700 cal. BP
     Final PPNB/PPNC   8600–8250 cal. BP
   Pottery Neolithic
     Yarmukian   8250–7800 cal. BP
     Wadi Raba   800–6500 cal. BP

Chalcolithic   4500–3800 BC

Early Bronze Age
   EB I   3800–3000 BC
   EB II   3000–2650 BC
   EB III   2650–2200 BC
   EB IV   2200–2000 BC
   (Intermediate Bronze Age)

Middle Bronze Age   2000–1550 BC
   (further subdivisions not indicated here)

Late Bronze Age   1500–1200 BC

Iron Age   1200–586 BC

Babylonian and Persian Periods   586–332 BC

Hellenistic Period   332–37 BC

Roman Period
   Early Roman   37 BC–132 AD
   Late Roman   132–324 AD

Byzantine Period   324–638 AD

Early Arab Period   638–1099 AD

considered to be the link between hunter-gatherers and the first farming communities. They are characterized partly by either semisedentary or fully sedentary hamlets or small villages in the Mediterranean zone with much more substantial architecture than in previous periods. Like their predecessors they primarily used microlithic chipped-stone tools, and the use of ground-stone grinding implements greatly increased compared to earlier periods.

There was an explosion of symbolic art in this period, clear cemeteries that imply territoriality, and larger group sizes.

Natufian economy was based on hunting medium-sized and small mammals, primarily gazelle, and the exploitation of wild cereals (Valla 1998a). Although previous Kebaran peoples also relied heavily on wild large-seeded grasses including wild wheat and barley (Piperno et al. 2004; Kislev, Nadel, and Carmi 1992), the significant increase in sickle blades with gloss has lead many to believe that the Natufians were the first people to cultivate wild cereals. The Early Natufian period began during the warming of the post LGM, but during the Middle and Final Natufian the return to cool dry conditions that characterized the Younger Dryas event took hold, encouraging new adaptations to a somewhat harsher environment.

The *Neolithic* in the Southern Levant lasted from circa 11,600 cal. BP to around 7000 cal. BP or circa 5000 BC. It is divided into Pre-Pottery Neolithic (PPN-) A, B, and C, and Pottery Neolithic (PN-) A (also known as Yarmukian) and B (also referred to as the Wadi Rabah culture) (Bar-Yosef 1998; Kuijt and Goring Morris 2002). The PPNA is the period with the first undisputed evidence for village settlements, and more extensive use of cereals, although scholars do not agree about the evidence for cultivation or domestication of either wheat or barley at this time (Colledge 1998; Hillman and Davis 1990; Kislev 1999; D. Zohary 1989). Hunting was the prime source of meat with gazelle being heavily exploited along with wild cattle, boar, equids, birds, and fish. The majority of sites of this period occur either within or adjacent to the Rift Valley (for example, Jericho, Netiv Hagdud, and Dhra). Residences consisted of oval-to-subcircular structures with grinding stones and storage facilities.

In the subsequent PPNB period (divided into Early, Middle, and Late) there was widespread expansion of the Neolithic lifestyle in both economic orientation and geographic extent. Some of the most important sites in the Levant include Jericho in Palestine, 'Ain Ghazal and Beidha in Jordan, Yiftahel and Kfar Hahoresh in Israel, and Tell Aswad in Syria. These are the first fully Neolithic societies with large settled villages, an economy based on clear domestication and farming of cereals and legumes, domestic sheep and goats, elaborate burial practices involving the removal and plastering of the skull, and substantial domestic architecture with the use of lime plaster in construction. It is also the first period in which ritual architecture appears, sometimes in the form of large upright slabs of stone (Kuijt and Goring-Morris

2002). In the Final PPN or PPNC phase, there was a general decrease in population throughout the region.

The Pre-Pottery Neolithic period coincided with the return to more humid and warmer climatic conditions that marked the beginning of the Holocene. However, the end of the PPNB and the PPNC appears to be very close to the timing of the "8.2KY event," a globally recognized cold episode that might also have triggered drier conditions in the Southern Levant. This abrupt change from the more favorable environment that existed throughout most of the PPNB might have led to considerable stress on the PPNB economic and social systems and be at least partially responsible for the shift to the PPNC and subsequently the PN periods (Bar-Yosef 2001, 2002b; Kuijt and Goring-Morris 2002).

The *Pottery Neolithic* is marked by the first use of fired pottery in the region. The earliest PN is identified as the Yarmukian, which is found throughout much of the Southern Levant and characterized by a mix of round and rectangular structures; faunal assemblages that include domesticated sheep, goat, bovids, and pigs; cereal and legume agriculture; and tool kits including sickle blades, arrowheads, and ground-stone tools for grinding and pounding (Gopher 1998). The following Lodian (or Jericho IX) division of the PN consists of a scaled-down occupation phase of several hundred years. Sites are small with scarce architectural remains, although mostly within the same territory as the Yarmukian. The Wadi Raba culture postdates the Lodian and is found throughout the Southern Levant. It has rectangular architecture including some multiroom structures. The Pottery Neolithic represents a shift to a more fully agricultural economy with diminished use of wild resources. According to Gopher (1998) this is a period that reflects a recovery from an economic and social crisis at the end of the seventh and beginning of the sixth millennia BC. The PN continues until the beginning of the fifth millennium BC.

A florescence of human settlement and culture takes place in the Southern Levant around 4500 BC, which marked the beginning of the *Chalcolithic period*. This was an outgrowth of successful adaptations in the Pottery Neolithic, accompanied by significant population growth and an expansion of villages across the Southern Levant (Gilead 1988, 1990). These villages appeared in a variety of different environmental zones throughout the Levant. This village network grew into a two-tiered hierarchy of settlement (T. Levy 1998a).

These settlements were supported by increasingly more intensive agropastoral economic systems including floodwater farming of cereals (Rosen and

Weiner 1994), olive and date horticulture (Zohary and Hopf 2000), and rais-
ing of domestic animals, primarily sheep and goats, with some pigs and cattle
as well. This is the period when experts suggest the "secondary products revolution" began to have a major impact on the Levantine economies. It is the
switch from the use of domesticated animals primarily as sources of meat to
their use as sources of milk products and fiber (Sherratt 1981; T. Levy 1992).
For the first time, settlements had associated cemeteries that were located outside the living area of the village. Large cemeteries were located at the sites
Kissufim, Azor, Shiqmim in the northern Negev, and Adeimeh near the site of
Tuleilat Ghassul in the lower Jordan Valley.

The Chalcolithic period is also notable for the development of craft specialization. This is especially true with copper metallurgy, primarily in the
production of ritual items, and also crafts undertaken by experts in specialized
ceramics, and stone carving. Some of the metal objects such as copper mace
heads were probably associated with ritual and public sanctuaries. There are
three such sanctuaries identified for the performance of public cult rituals at
the sites of Tuleilat Ghassul, Ein Gedi near the Dead Sea, and Gilat in the
northern Negev Desert (Gilead 1988; T. Levy 1998a).

Around 3800 BC, the network of Chalcolithic villages suffered a collapse
and general population decline with a breakdown in the settlement hierarchy
and a return to small autonomous villages that for the most part characterized
the subsequent Early Bronze Age I period. Most experts suggest a combination
of social and environmental factors that led to the decline of the Chalcolithic
traditions (T. Levy 1998a; Joffe 1991).

The climatic picture during most of the Chalcolithic period is characterized by significantly more rainfall than the region receives in the present day
(see chapter 4) (Bar-Matthews and Ayalon 2004). At the end of this period,
around the second half of the fourth millennium BC, the rainfall dropped to
below that of the present. This would have been a point of environmental
stress to the Chalcolithic economic system, which might have been over-
stretched due to resource procurement for elite consumption and a growing
sense of territoriality on the part of Chalcolithic communities (Joffe 1991).

Village life returned and developed in the *Early Bronze* Ia beginning around
3800 BC. By the EB Ib period, the first urban centers developed in the Southern
Levant at the sites of Tel Erani in southern Israel and Megiddo in the north.
These two sites were both stimulated by some form of influence from Egypt, although the exact nature of this contact is still unclear. Some scholars believe that

the proto-Canaanite settlements were trading colonies under the control of Egyptian entities (Richard 2003). Trade was a significant factor in relations with Egypt, with agricultural products such as olive oil and wine among the primary commodities being exported from the Levant. The population expanded again at this time, and there was a revival of craft specialization. This seems to have stimulated wealth and centralization of power among the urban elites throughout the EB II period beginning about 3000 BC. By this time any possible Egyptian political control over the Southern Levant had diminished, and there were a number of fully urbanized independent settlements.

Towns such as Tel Yarmouth in the central Shephela foothill zone boasted large well-constructed palaces and temples and massive fortification walls. The equally impressive site of Arad in the Negev Desert appears to have controlled the metal trade from the Sinai. Although initially stimulated by Egyptian influence, Southern Levantine urbanism was very much an outgrowth of local culture. By the EB II there was a uniformity of culture in the form of architecture, ceramics, and cultic traditions. The links between palaces and temples at sites such at Yarmouth seems to suggest that a theocratic elite were controllers of political and economic systems in the region at this time (de Miroschedji 1988). The end of the EB II was marked by some abandonments of sites such as Arad. Although there are climatic fluctuations throughout the Early Bronze Age, these are little different from the centuries before and after the EB II. The abandonment of Arad seems to have more to do with the politically powerful Third Dynasty coming to power in Egypt and co-opting the mineral resources in Sinai (Richard 2003).

The EB III period beginning around 2650 BC saw a renewal of the outward manifestations of political strength. New building works on fortifications, temples, and palaces occurred at a number of sites in the region, suggesting a return to a powerful ruling elite and a strong urban society. Trade connections with Egypt had ceased, and the economic base seems to have come from control over wine and oil commodities for internal markets (Joffe 1993). External connections now came from the north rather than the south. However, by the end of this period at circa 2200 BC, most of the urban centers in the Southern Levant had been abandoned. The reasons for this are likely to be in part climatic since this corresponds with the so-called 4.2KY event marked by droughts in the Near East, but also in part a built-in internal weakness in the social and economic structure of the society. These events are covered in detail in chapter 6.

The subsequent period, known as the EB IV (it also appears in the literature as the Intermediate Bronze Age, EB-MB, or MB I), was, for the most part,
marked by a much-reduced population demographic. The remaining population in much of the Southern Levant was organized into autonomous agro-
pastoral villages or semisedentary pastoral camps. These are often found in
the vicinity of springs. There are, however some notable exceptions to this settlement pattern, primarily in Transjordan, where a handful of large urban
centers such as the sites of Khirbet Iskander and Iktanu carried on with large
populations and fortified towns (Prag 1989; Richard 1990). It is notable that
these two sites are located near perennial streams that might have been a dependable source of water for irrigation farming (A. Rosen 1995).

The following *Middle Bronze Age* (MB) beginning circa 2000/1900 BC is a
cultural entity that clearly differs from the preceding Early Bronze Age. Some
of the most important new traits of the period include the use of true bronze
artifacts as opposed to the copper artifacts that were common in the EB, the
introduction of the fast wheel for pottery manufacturing, massive town fortifications that included large earthen ramparts, renewal of monumental building, and an extensive hierarchy of settlements from villages to cities (Ilan
2003). Analyses of burial populations from MB tombs and graves indicate an
influx of new populations, probably from more northern regions, bringing
cultural links with Syria. The period began with small villages that later developed into walled towns, initially on the coastal plain of Israel and in the in-
terior valleys. With time, new walled settlements expanded into the central
hills, and by the time of the MB II–III, the new society had expanded throughout the Southern Levant as far south as the northern portion of the Negev
Desert. However, most of Transjordan was untouched by these developments
(Ilan 2003).

The period has a distinct militaristic persona with the renewal of massive
fortifications and the adoption of new weapon types such as bronze daggers,
lance heads, and chariots. A new and powerful elite class developed as indicated by palaces, wealthy tombs, and rich hordes of jewelry. The society
showed every sign of well-established social hierarchies with rulers, elite
classes, priests, and full-time craftspeople.

The end of the Middle Bronze Age is marked by a long and gradual series
of town destructions and abandonments eventually resulting in the transformation of the remaining settlements into the Late Bronze Age communities
rather than a sharp change in settlement patterns or material culture. How-

ever, societies that remained retained their urban character. Many scholars mark the end of the MB by the historical events relating to the expulsion from Egypt of the "foreign" Hyksos rulers circa 1540 BC. This roughly corresponded with a period of about 100 years in which some towns and villages yielded remains of violent destructions and abandonments (Ilan 2003). Other scholars suggest that a large part of the collapse and transformation of MB settlements might be related just as much to internal strife as to external invasions and attacks (Bunimovitz 1998).

The Middle Bronze Age is clearly the most developed complex society the region had yet seen. However, most interestingly for the thesis of this volume, this highly developed complexity, wealth, and large population growth all took place under climatic conditions that were significantly drier than those at the present time. One can conjecture that this was made possible by new technologies for rock-cut wells, cisterns, and perhaps canal-based irrigation systems.

The *Late Bronze Age* (LB), beginning circa 1500 BC, was characterized by a new territoriality of discrete polities, many of which were vassal states under the control of the Egyptian hegemony. Much of our knowledge of the politics of this period comes from a set of inscriptions found in Egypt, known as the Amarna Letters. These were written by vassal Canaanite kings sometimes pleading for military assistance from Egypt to help defend their towns from nomadic and other incursions. The richness of this written record has added a different dimension to our understanding of the history of the region, but in some ways it has inhibited our general understanding of Late Bronze Age society due to the overemphasis some scholars have placed on these reported events (Bunimovitz 1998). As far as we can see from settlement patterns of sites dating to this period, societies maintained an urban character, but fortifications were generally left in disrepair. The sphere of influence of these smaller urban polities was likewise reduced. Bienkowski (1989) suggests that in spite of some examples of palaces, temples, and rich grave goods, for the most part these communities were under economic stress due to the burden of tribute payments to Egypt. Knapp (1992) characterizes the Late Bronze Age towns as less significant nodes along much larger systems of world trade routes. In terms of environmental conditions, the LB and the MB were very much alike with rainfall amounts roughly equivalent to or less than the levels of the present day.

The transition from the LB to the beginning of the *Iron Age*, Iron Ia period (ca. 1200 BC) is marked by a wave of destructions of LB Canaanite towns and

the appearance of new sociopolitical and ethnic elements that appear in historical records as Philistines, Israelites, Ammonites, Moabites, and Edomites
(see the brief but comprehensive summaries in Dever 1998; Finkelstein 1998;
Holladay 1998; LaBianca and Younker 1998; Stager 1998; and Younker 2003).
The result was a territorial division along ethnic and sociopolitical lines into
Philistines on the southern coast, Canaanites dominated by Egyptian control
in the central and northern coastal plain and inland valleys, and more loosely
organized agro-pastoral groups occupying the Southern Levantine highlands
and the Jordan Valley. The Canaanite cities came under siege during the inva-
sion of the Sea Peoples (later partially identified with the Philistines) by the
Egyptians, who had been their former protectors, as well as by the newly
emerging Israelite polity in the highlands.

The Iron Ib period from circa 1150 to 1050 BC is marked by the decline of
Egyptian control over Canaan, the consolidation and expansion of the
Philistines, and an expansion and growth of settlements in the highlands. The
highland settlements, equated with the Israelite ethnic groups, were characterized by evidence for agricultural intensification in the form of hillslope terrace systems, water collection technology via cisterns, storage pits, and silos.
There was also expansion of these groups into new areas such as the Galilee.

Biblical references to the territorial monarchies of Israel, Ammon, Moab,
and Edom have led to much archaeological discussion and debate about the
historical events of the Iron Age II period. Excavations at large sites such as
Hazor, Gezer, and Megiddo have revealed massive fortifications including city
walls with a new kind of partitioning called casemate, massive six-chambered
city gates, and other examples of public architecture at these Israelite sites. Israelite fortresses also appear on the frontier borders in the Negev Desert.

The Iron IIb period is associated with historical events that led to the division of the Israelite monarchy into two kingdoms, Israel in the north and Judah in the south. This period is associated with a multitiered settlement
hierarchy. The Iron IIc represents the period after the conquest of the northern part of the region by Assyria, with growth in towns of the southern polity
of Judah as well as Transjordan.

The climate of this period appears to have been very similar to that of the
preceding MB and LB periods, with rainfall conditions that seem to have been
drier than those at present. Again, it is significant that the social and political
events surrounding the development of the first secondary state in this region
of the Levant took place under a rainfall regime that was even less beneficial

for dry faming than that of the present (Bar-Matthews and Ayalon 2004). This may have much to do with the improvements in water management systems as demonstrated by the evidence for rock-cut cisterns, wells, and hillslope terrace systems. This intensification of agricultural systems would have allowed higher and more predictable yields of both subsistence crops such as cereals and cash crops such as olives and grapes.

In the *Persian period*, the Southern Levant once again came under the rule and influence of a foreign power. The beginning of this period is dated from the Babylonian conquest in 586 BC, and the rise of Persian control in the region. However, there were still significant territorial divisions throughout this period that were manifested by three political and ethnic entities including (1) the ethnic polities of the highlands and Transjordan such as Judah, Samaria, Ammon and Moab, which were strongly influenced by the Persian cultural sphere, (2) the Phoenician and Greek commercial centers along the coastal plain, and (3) the Arab populations who controlled the south. The communities on the coastal plain were economically orientated toward the west with very close ties to Greece. There was a large growth in international trade at this time, and there is a great deal of evidence for imports of commodities from Greece, including abundant quantities of Greek wine. This coastal population encouraged rapid Hellenization even predating the Greek conquest at the beginning of the Hellenistic period in 332 BC (Stern 1998).

In the *Hellenistic period*, the Greek Ptolemies bolstered the economy of the coastal cities through increased trade, while the Nabateans controlled overland trade from the east through the southern trade routes across the Negev. They founded many of the Negev Desert sites that later became important Byzantine period cities, such as Nessana, Elusa, and Oboda. The hill country during the third century BC was sparsely populated. Populations of these regions increased in the later portions of the Hellenistic period, which continued until 63 BC, when the Roman hegemony expanded to include most of the Southern Levant (Berlin 2003). The Roman and Byzantine periods in this region are discussed in detail in chapter 7 and therefore are not covered in this brief background summary. The climate during the Persian, Hellenistic, and early part of the Roman periods appears to have been somewhat moister than previous periods, and roughly similar to that of the present (Bar-Matthews and Ayalon 2004).

Paleoenvironments of the Near East: The Retreat of the Pleistocene

The Pleistocene began roughly 2.5 million years ago and was characterized by major shifts in global climate between glacial and interglacial episodes. It continued until circa 11,500 years ago with the commencement of the current Holocene warm interglacial. Together, the Pleistocene and Holocene are known as the Quaternary period. The Pleistocene was first defined by nineteenth-century European geologists who identified and described the ancient Alpine moraines and glacial *erratics* (giant individual boulders located far from the source rock) in continental Europe. Later geologists used sequences of glacial till deposits to define stages of terrestrial glaciation that they named according to local place names. Therefore, the last glacial episode is referred to variously as the Würm in Alpine terminology, Weichselian in Northern Europe, Devensian in Great Britain, and Wisconsinian in North America (Lowe and Walker 1997:11). In older paleoenvironmental literature concerned with the Near East, paleoclimatologists have traditionally used the Alpine terminology (Butzer 1975). From the 1950s onward, analyses of oxygen isotopes from sea cores demonstrated that most terrestrial glacial sequences were oversimplifications of glacial advances and retreats (Emiliani 1955; Shackleton 1967; Pisias et al. 1984). Studies of oxygen isotopes from the $CaCO_3$ tests of microscopic marine animals from around the world, and from water molecules in ice cores from polar regions, produced a long record of temperature changes that spanned hundreds of thousands of years, eventually including all of the Quaternary period. Allowing for variations in sedimentation and salinity, these records are remarkably consistent in their results due to the close relationship between $\delta^{18}O$ concentrations in the atmosphere, oceans, and ice. These sequences of cold versus warm episodes are represented by curves showing numerous peaks and lows (see figure 2.3) that are divided up into stages. The peaks indicate warm phases and are designated by odd numbers; the lows identify cold periods that are signified by even numbers.

Before beginning a discussion of Holocene paleoclimates, it is important to understand the major climatic changes that took place at the end of the Pleistocene. These led to environmental fluctuations that coincided with the Natufian period in the Levant and have been cited by some researchers as playing a major role in the shift from broad-spectrum hunting and gathering to early plant cultivation (see the discussion in chapter 6). The most profound environmental shifts of the Late and Terminal Pleistocene were the warming trend occurring just after the Late Glacial Maximum (LGM), the cold dry reversal of the Younger Dryas event in the Terminal Pleistocene, and subsequently the Pleistocene-Holocene transition to warmer wetter climatic conditions.

The Late Glacial Maximum occurring between 25,000 and 18,000 cal. BP (Roberts 1998) represented the final maximum extent of the continental ice sheets, as indicated by Stage 2 on the oxygen isotope sequence. After this episode global warming ensued until the period designated the Younger Dryas from circa 12,700 to 11,500 calendar years BP (after Cronin 1999:205–206). Archaeologically this period coincided with the Epipaleolithic Natufian period from circa 15,000 to 11,600 calendar years BP (Byrd 1994; Bar-Yosef 2002a).

The Younger Dryas was first identified from plant macrofossils in Europe (Jensen 1935; Iversen 1954). It represented a distinct return to cold dry–loving vegetation, after an initial warming phase at the end of the Late Glacial Maximum. This cold dry episode has since been identified in pollen diagrams from Europe, in some parts of Asia and North America, as well as from South America and Africa (Cronin 1999:205). Evidence for it appears in both marine and ice cores from a number of localities worldwide (Roberts 1998:72–76). Although perhaps not completely a global event (see Cronin 1999:205–209), the Younger Dryas was a widespread cooling episode that had a significant impact on vegetation, landscape, and water resources in the Near East, an effect that is outlined below.

The shift back to a warming trend that ushered in the Holocene was likewise responsible for major changes in vegetation complexes, landscape, and hydrology. The evidence for these ecological shifts is summarized below.

OXYGEN ISOTOPE EVIDENCE

Probably the most powerful tool for understanding climatic change is the proxy evidence from the oxygen isotope record. For the Near East we have recent oxygen isotope sequences from lake cores, marine cores from the Mediterranean Sea (Kallel et al. 1997; Kroon et al. 1998; Schilman et al. 2002), and oxygen isotope determinations from speleothems in caves from the Galilee (Geyh 1994), central Israel (Bar-Matthews, Ayalon, and Kaufman 1997, 1998; Bar-Matthews et al. 1999; Bar-Matthews and Ayalon 2004), and the Judean Hills (Frumkin, Ford, and Schwarcz 1999).

One of the best isotopic records in the Levant comes from the speleothems of the Soreq Caves (Bar-Matthews, Ayalon, and Kaufman 1997, 1998; Bar-Matthews et al. 1999; Bar-Matthews and Ayalon 2004) (see figure 4.1). These caves are located in central Israel in an area that today receives a mean precipitation of 500–550 mm per year. Present-day mean temperatures within the cave are 18°C–20°C. This isotope record has been dated by ^{230}Th/^{234}U thermal ionization mass spectrometry (TIMS) and represents the past 60,000 years of the Late Pleistocene and Holocene. In this study Bar-Matthews and others measured isotope ratios for oxygen (δ^{18}O) and carbon (δ^{13}C) from twenty different speleothems. The δ^{18}O record clearly shows a major peak for the Late Glacial Maximum (LGM) at about 19,000 years ago indicating extremely cold climatic conditions. After this there is a relatively steep decline punctuated by several sharp fluctuations that reach a low point at circa 14,000 years BP. The rise in δ^{18}O reaches a plateau from circa 13,200 to 11,400 years BP and is associated with the Younger Dryas cooling event. This pattern is repeated in the curve for δ^{13}C, which also rises to a plateau during this time period, although it is not as marked a rise as that of the δ^{18}O record. An 1800-year Younger Dryas is a very long duration compared to other records of the Younger Dryas both within the Levant and in other regions of the world where the YD lasted approximately 1000–1300 years.

Bar-Matthews, Ayalon, and Kaufman (1997) developed a model from these data for estimating mean annual rainfall. The model relies on the modern relationship between the average annual rainfall amount, its δ^{18}O composition, and δ^{18}O values for the cave water. The results of these analyses indicate that the δ^{18}O peak at 19,000 years BP (the LGM) was the coldest ($\sim$12°C) and driest ($\sim$250 mm rainfall per year) episode since that time period (Bar-Matthews, Ayalon, and Kaufman 1997; Bar-Matthews et al. 1999). The correspondingly high δ^{13}C value at 19,000 years BP is also considered to be cli-

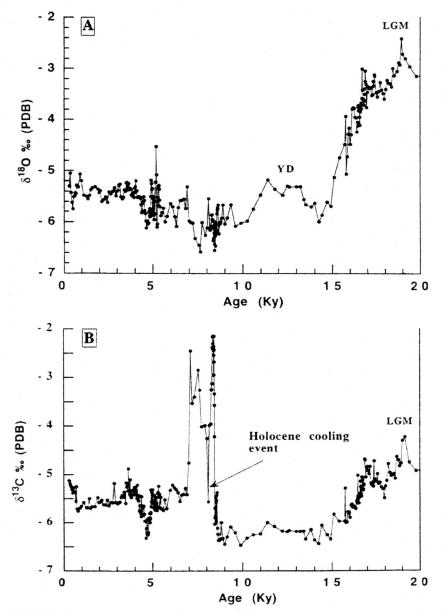

FIGURE 4.1
Oxygen and Carbon Isotope Curves from Soreq Cave (After Bar-Matthews et al. 1999).

matically controlled, and Bar-Matthews and others (1997; 1999) interpret this as indicating the mixing of C_3 and C_4 vegetation with a dominance of C_4 types, which are typical of colder and drier climatic conditions. Subsequent to the period of the LGM, drops in both the curves are interpreted by Bar-Matthews et al. (1999) as indicating warming conditions accompanied by a gradual increase in rainfall. The section of the isotope curves corresponding to the Younger Dryas indicates not only cool but also dry climatic conditions, although these would not have been as pronounced as in the LGM.

Geyh (1994) also presents isotope data first published by Issar (Issar 1990; Issar et al. 1992) from nineteen stalagmites collected from six caves in the Galilee. These are dated by a large number of radiocarbon determinations. Geyh warns against assuming a simple relationship between these $\delta^{18}O$ and $\delta^{13}C$ data and temperature change, due to possible local kinetic factors that could influence the isotope fractionation. The sequence he presents illustrates isotopic peaks during the Pleistocene, as in the Soreq Cave data, with a steep drop-off at the Pleistocene-Holocene boundary. The Galilee caves also show a series of peaks defining the YD, in this case from circa 13,000 to 11,500 cal. BP. In both the Soreq and the Galilee caves the YD is represented by two or more cold dry peaks rather than a single cold dry phase. The Galilee record indicates fluctuating climatic conditions with a major increase in warmth and humidity at the close of the Pleistocene. The Soreq record suggests that a major increase in warmth and humidity preceded the YD, beginning at circa 17,000 BP, a couple thousand years after the LGM.

Frumkin, Ford, and Schwarcz (1999) examined a sequence of $\delta^{18}O$ values from speleothems in three caves from the Judean Hills near Jerusalem. The results provided a long (ca. 150,000-year) record of climate changes in this region. Although the emphasis is on the Pleistocene paleoclimatic history and its periodicity, and the Terminal Pleistocene–Early Holocene dates are somewhat out of phase with those from the Soreq caves, Frumkin, Ford, and Schwarcz (1999) point out a general agreement between the results of their analyses and the Soreq cave 2-N. There is a similar decline in $\delta^{18}O$ from its high peak marking the cold dry climate of the LGM down to the beginning of the warmer and wetter Holocene. This decline begins shortly after the LGM in agreement with the Soreq data and in opposition to the Galilee cave spectrum. As in both records this decline is interrupted by a return to cool dry conditions marking the Younger Dryas event.

The isotope data constitute the most important proxy we have to date from which to estimate temperature and precipitation in the past. The isotope spectra from speleothems in Levantine caves discussed above give us a picture of terminal Pleistocene climate, which began with extreme cold and dry conditions at the Late Glacial Maximum, circa 19,000–23,000 cal. BP. Shortly afterward there was a rapid change to a warmer moister environment over the course of some 4000 years. At the end of this time period the Younger Dryas event, from circa 13,000 to 11,500 cal. BP, marked the renewal of cold dry conditions in a short-term climatic reversal. The YD ended with a return to a warm moist setting that ushered in the dawn of the Holocene. Given this isotopic evidence for climatic change we can move on to other forms of proxy information that provide clues to other aspects of the Levantine environment, including vegetation and landscape.

PALYNOLOGICAL EVIDENCE

There have been numerous diagrams made of pollen taken from lakes and marine cores throughout the Eastern Mediterranean regions of Greece, Turkey, Syria, and Israel (see, for example, Baruch 1994; Bottema, Entjes-Neisborg, and van Zeist 1990; El-Moslimany 1982; Niklewski and van Zeist 1970; Roberts, Meadows, and Dodson 2001; van Zeist and Bottema 1982). Because of problems with dating and differences in sediment deposition rates it has been a difficult task to combine this information into a regionally coherent picture of climate and vegetation change in the Late Glacial and Early Holocene. Rossignol-Strick (1995, 1997) has offered one of the most successful attempts at articulating these diagrams. In examining the pollen records from a number of cores in the Mediterranean and Arabian Seas, and terrestrial lakes in Greece, Syria, Turkey, Israel, and western Iran, she described two main mutually exclusive pollen phases. The first is the *Chenopodiaceae phase*, which she correlates with the cold dry Younger Dryas from circa 12,700 to 11,500 cal. BP; the second is the *Pistacia phase*, which she associates with the Post-Glacial–Early Holocene warm moist episode occurring from circa 10,000 to 7000 cal. BP. Rossignol-Strick identifies these two phases in cores from the Adriatic, the Tyrrhenian, and the Ionian seas, as well as from several lakes in Greece, the Ghab Valley in northwest Syria, Lake Hula in northern Israel, Lake Van in eastern Turkey, and cores from Lake Zeribar in western Iran.

The *Pistacia* and Chenopod phases occur to varying degrees in all of these diagrams, but their timing is based primarily on radiocarbon dating and is often out of phase. Rossignol-Strick suggests that the dates should be adjusted so that all of these pollen diagrams display the above two phases within similar time ranges. Although this approach is a major step forward in our understanding of vegetation responses to regional climatic change across the Near East, she doesn't account for the important aspect of localized environmental factors around the terrestrial core sites and the locations of Pleistocene plant refuges. Such factors might offset the timing and degree of plant colonization.

Two important lake and marsh settings in the Levant have provided the most information on the Younger Dryas and Pleistocene-Holocene transitions for that region. These are Ghab in northwestern Syria and Lake Hula in northern Israel. The interpretation of the pollen sequences from these two regions located respectively in the Northern and Southern Levant has been the subject of much debate among paleoclimatologists and archaeologists, especially with respect to the period of the Pleistocene-Holocene transition, where the pollen percentages are markedly out of phase between these two regions. This is unlikely to be a result of extreme differences in climate due to the close correlation of pollen curves from earlier portions of the cores (Weinstein-Evron 1990). The lack of correlation for this critical time period is most likely due either to problems in dating or the impact of humans on the environment that might have altered the pollen signal. The key role of these two lakes in reconstructing environmental and climatic change in the Levant at this critical juncture merits the detailed discussion of the following pages.

Lake Ghab

The Ghab Valley is located in northwest Syria at the northern end of the Syrio-African Rift. The valley is bordered on the west by the Alouite Mountains, which rise up to 1700 m, and on the east by the Zawiye Mountains, which are only about 800 m in elevation. As with the Hula basin, precipitation varies a great deal due to the variation in topography and geographic zones of the surrounding mountains. Therefore rainfall varies from about 1300 mm over the Alouite Mountains down to 700 mm in the valley itself, and 600 mm on the western flank of the Zawiye Mountains to the east (Baruch and Bottema 1991). Three vegetation zones at different altitudes in the region are described. These are the *Ceratonia-siliqua–Pistacia lentiscus* forest up to 300 m,

Quercus calliprinos–Pistacia palaestina forest up to 800 m, and the *Quercus infectoria* forest above 800 m (Niklewski and van Zeist 1970).

Two teams of researchers produced pollen sequences from the marshland in Ghab. The first team was Niklewski and van Zeist (1970), and much later Yasuda, Kitagawa, and Nakagawa (2000) recored the valley. Niklewski and van Zeist extracted the pollen cores in three sections, two from the southern portion of the lake and one from the north. These have been combined to form one sequence (see figure 4.2). The relevant diagram for this discussion is the Ghab I core. This core is dated by three radiocarbon dates, two of which are close to the upper range of radiocarbon at circa 47,000 bp (at 650 cm) and therefore not within the scope of this discussion. These older dates, however, were used to calculate the rate of deposition of sediments in the lake and therefore useful for interpolation of dates throughout the core. The third radiocarbon determination on shell fragments at about 133 cm was circa 11,450 cal. BP (10,080 ± 55 bp) (Niklewski and van Zeist 1970). There has been much discussion in the literature about the apparent inconsistencies between the Ghab and the Hula diagrams (see, for example, Butzer 1975; Baruch and Bottema 1991; Moore and Hillman 1992; Baruch 1994; Rossignol-Strick 1995). Authors such as Rossignol-Strick (1995) and Cappers, Bottema, and Woldring (1998) suggest problems with radiocarbon dates in one or the other lake basin, or alternatively a major zonation in climatic regimes between the Southern and Northern Levant that is not in existence in modern times (Butzer 1975). The uncorrected sequence suggests that during the cold dry episode of the LGM, there was an expansion of the Mediterranean forest into the Ghab basin (corresponding to subzone Y_1, dated by extrapolation to circa 25,000–23,100 cal. BP, and subzone Y_2–Y_4, circa 23,500–17,000 cal. BP, extrapolated). Following this, with the global warming that ensued after the LGM, the forests in the Ghab receded to their minimum extent represented by subzone Y_5 (ca. 17,000–13,000 cal. BP, extrapolated). The forest then recovered with a dramatic rise during subzone Z_1, from circa 13,500 (extrapolated) to 11,450 cal. BP (10,080 ± 55 bp), a time period that would correspond to the Younger Dryas event. This contrasts with the Hula diagram described below, which shows major forest advancement in the Galilee after circa 17,500 corrected and calibrated BP, and significant forest recession during the Younger Dryas from circa 13,000 cal. BP.

Most of this interpretation of the Ghab diagram is based on extrapolated dates using an endpoint of two unreliable radiocarbon dates near the bottom

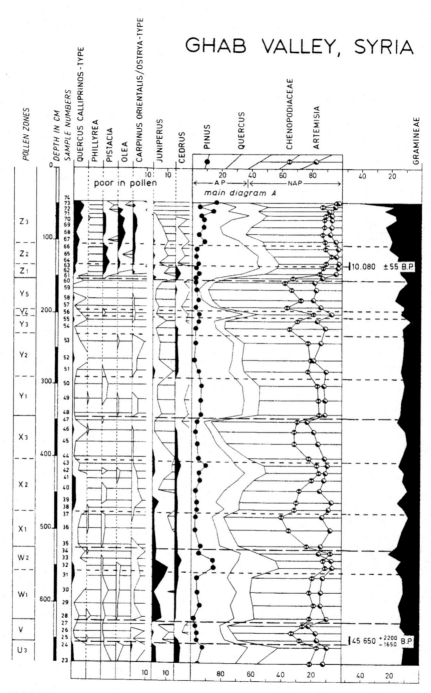

FIGURE 4.2
Niklewski's and van Zeist's (1970) Pollen Diagram from Lake Ghab.

of the core. It seems more reasonable in this case to compare the major pollen zones with those from other diagrams and readjust the chronology accordingly. This approach was taken by Rossignol-Strick (1995), who identified the *Pistacia* phase dated elsewhere from circa 10,000 to 7000 cal. BP (9000–6,000 bp) and the *Chenopodiaceae* phase, which is associated with the Younger Dryas, in other cores from circa 13,000 to 11,500 cal. BP, as is discussed above. This allows the major forest decline and increase in chenopods in zone Y_5 to correspond with the Younger Dryas. This interpretation requires an adjustment to the original radiocarbon date of 10080 ± 55 bp, pushing it later to circa 9800 cal. BP (8700 bp); however, it puts the pollen sequences in line with the other pollen data from the region.

Yasuda, Kitagawa, and Nakagawa (2000) were able to obtain nine ^{14}C dates from their core, exclusively on freshwater mollusks. They suggest that unlike the situation at Lake Hula, the Ghab core requires no correction for the hard water effect. They base this on their chenopod/*Artemisia* ratio, which they claim indicates cool dry conditions at a depth of 470 to 430 cm, dating between circa 13,500 and 11,500 cal. BP, a time corresponding to the Younger Dryas event (see figure 4.3). Wright and Thorpe (2003:57) dispute this claim. They suggest that a chenopod rise in relation to *Artemisia* more likely indicates a local expansion of chenopods due to the exposure of saline soils during a period of low lake level. They support this interpretation with evidence from Lakes Van in Turkey and Zeribar in Iran, where chenopods and *Artemisia* curves are correlated. Additionally, the 470–430 zone contains high percentages of *Quercus* (oak) pollen, which seems to preclude it from representing the Younger Dryas event. Wright and Thorpe (2003) point out that the corresponding zone in the Niklewski and van Zeist (1970) core contains a distinct rise in *Pistacia*, which would indicate that it should be assigned to the Early Holocene. They also suggest that the underlying pollen zone at 590–470 cm has a more pronounced peak for both *Artemisia* and chenopods and is therefore a better candidate for the Younger Dryas event. If this is indeed the case, then the ^{14}C dates are too old and should be adjusted by the same order of magnitude as that suggested by Rossignol-Strick for the Niklewski and van Zeist (1970) core.

Lake Hula

Lake Hula is located in the northern Galilee of Israel within the Syrio-African Rift Valley. It is bordered on the west by the hills of the Upper Galilee

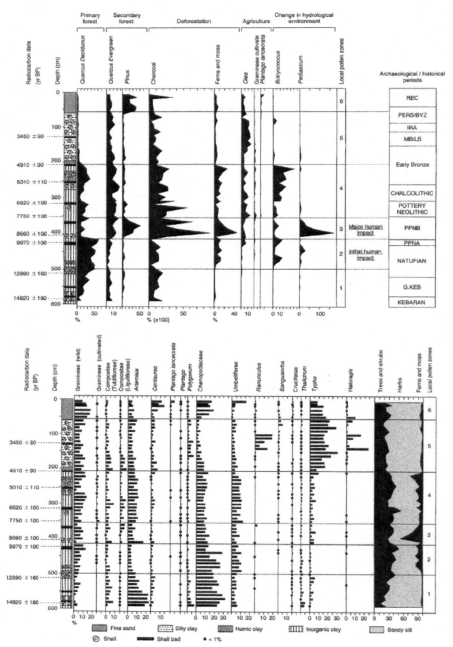

FIGURE 4.3
Yasuda, Kitagawa, and Nakagawa's (2000) Pollen Diagram from Lake Ghab.

and on the east by Mt. Hermon and the Golan Heights. Today the Hula Valley receives about 450 mm of rainfall per annum but also benefits from runoff from the Hermon, which receives up to 1500 mm per annum and about 1000–500 mm per year from the upper Galilee and Golan Heights. The valley today is typified by *Quercus ithaburensis* (oak) park forest, which includes *Pistacia atlantica* (pistachio) and *Amygdalus korschinskii* (almond) (M. Zohary 1980).

The most recent core from Lake Hula (Baruch 1994; Baruch and Bottema 1991, 1999) has a number of radiocarbon dates that bracket the Late Glacial Maximum, fix the period of the Younger Dryas, and extend throughout the Holocene (see figure 4.4). If we examine the total arboreal pollen percentages versus those for nonarboreal pollen, a simple story is told in which there are very low tree pollen percentages at the end of the LGM, dated in this core at circa 20,500 cal. BP (17,140 ± 220 bp). This represents the cold arid phase of the Pleistocene Pleniglacial. A change occurs at circa 18,700 cal. BP (15,580 ± 220 bp) that is marked by a rise in arboreal pollen culminating in a peak of over 75% at circa 13,550 cal. BP (11,540 ± 100 bp). There is an ensuing sharp drop in the tree pollen percentages, down to a low of less than 25% at a depth of 1140 cm, at an extrapolated date of 12,500 cal. BP, which Baruch and Bottema (1991) and Moore and Hillman (1992) equate with the YD. Tree pollen rises again reaching a modest peak of 45% at circa 12,400 cal. BP (10,440 ± 120 bp) marking a return to moister and warmer climatic conditions shortly before the beginning of the Holocene.

Rossignol-Strick (1995) presents a different interpretation of this diagram. As with Ghab, instead of pinpointing the YD on the diagram by the radiocarbon dates, she determines its location with reference to the percentages of the key plant types. Rossignol-Strick maintains that during the time period of the YD as designated by the radiocarbon dates there is a rise (albeit slight) in *Pistacia*, a tree normally associated with warmer wetter climatic conditions. Based on this observation, she assigns the YD to the lower portion of the core, which is typified by minimal to no *Pistacia*, very low percentages of *Quercus*, and high percentages of chenopods and *Artemisia*, both indicators of dry climatic conditions. Although this interpretation makes sense from a broad multiregional perspective, it requires us to radically readjust the four lowermost radiocarbon dates of the core, thus making them younger by around 4000 years. Rossignol-Strick (1995) suggests that there might be a problem with radiocarbon dates in general in Lake Hula due to possible influx of old carbonates from hard water. Cappers, Bottema, and

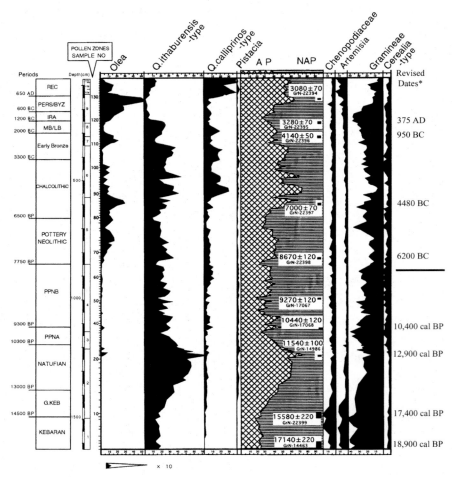

FIGURE 4.4
Pollen Diagram from Lake Hula (After Baruch and Bottema 1999) with Revised Chronology along Right Margin (After Cappers, Bottema, and Woldring 1998; and Wright and Thorpe 2003).

Woldring (1998) propose a correction factor of about 900 years younger for the Hula dates in the lower part of the core covering the Late Pleistocene and Holocene transition. Meadows (2005) examined the carbon chemistry of Lakes Ghab and Hula and suggests the even greater correction of about 5000 years for the lower Late Pleistocene–Early Holocene portion of the Hula core. The only thing clear about the dating of the Hula core is that scholars are in strong disagreement about the extent to which the Hula dates should be adjusted, if at all.

Moore and Hillman (1992) proposed a close correlation between the Hula core and the well-dated macrobotanical remains from the site of Abu Hureyra on the Syrian Euphrates. Although the site of Abu Hureyra has always been located in the steppe zone of northern Syria, the use of Mediterranean oak-Rosaceae forest plants in the early occupation of the site (occupation period 1A ca. 13,600–13,000 cal. BP) indicates that the forest zone was closer to the site than in later periods. Further evidence from the remains of wild einkorn wheat and wild rye suggest that occupation period 1A at Abu Hureyra was characterized by relatively moist springs and/or summers. After this moist episode there was an abrupt change in the plants recovered from Abu Hureyra, which Moore and Hillman interpret as the beginning of a very dry episode in period 1B. This period is dated from circa 13,000 to 12,200 cal. BP and corresponds to Baruch and Bottema's (1991) decrease of arboreal pollen from circa 13,500 to 12,500 cal. BP, which is interpreted as the Younger Dryas. Moore and Hillman estimate that the abrupt shift to more arid zone plant resources took place around 12,500 cal. BP, becoming more pronounced in period 1C (ca. 12,200 to 11,500 cal. BP) (Moore and Hillman 1992).

Wright and Thorpe (2003) propose what appears to be the most reasonable interpretation of the Hula diagram to date, based on the Cappers, Bottema, and Woldring (1998) ^{14}C date correction and calibration to calendar years BP. This scenario is probably the best estimate until the core can be redated using terrestrial organic matter uninfluenced by the old carbon effect or using an alternative means of dating (see table 4.1). The base of the sequence begins with low arboreal pollen and high Chenopodiaceae and *Artemisia* from 1560 to 1430 cm, beginning at around 19,000 BP (corrected and calibrated). This marks the height of the Late Glacial Maximum, a time when cool dry conditions prevailed in the Near East. The high percentages of grasses, however, suggest moister, more temperate conditions than in other localities further inland in Anatolia and the Zagros. This is possibly due to the mitigating effect of the Mediterranean Sea. Oak is also more abundant than in other regions, suggesting proximity to the Pleistocene refuge of these trees.

Immediately after this Chenopodiaceae phase, with a change in climatic conditions after the Late Glacial Maximum at approximately 17,400 BP (corrected and calibrated), the forest zone in the Levant began to steadily expand, as indicated by the arboreal pollen at a depth of about 1475 cm. The forest was

Table 4.1.   Range of Recalculated Dates after Wright and Thorpe (2003).

Coring Site Date	Depth (cm)	Raw Date	Corrected Date	Calibrated
Ghab Marsh (Niklewski and van Zeist 1970)				
	129–137	10,080		11,353
	169–171	23,030		28,300
	645–655	45,650		ca. 50,000
Ghab Marsh (Yasuda, Kitagawa, and Nakagawa 2000)				
	130	3450 ± 90		3691
	210	4910 ± 90		5644
	252	5010 ± 110		5736
	315	6620 ± 100		7502
	352	7750 ± 100		8464
	405	8680 ± 100		9613
	425	9970 ± 100		11,187
	510	12,890 ± 160		15,259
	590	14,820 ± 180		17,723
Lake Hula				
	1120–1140	10,440 ± 120	9400	10,372
	1235–1242	11,540 ± 100	10,960	12,879
	1475–1500	15,560 ± 220	14,540	17,415
	1600–1625	17,140 ± 220	16,040	18,915

dominated by typical Mediterranean trees, primarily oak (*Quercus*), and included pistachio (*Pistacia*). The appearance of pistachio is a strong indicator of warmer and moister conditions. The pollen indicating this forest expansion reaches a peak at about 1240 cm, dating to circa 12,879 BP (corrected and calibrated) and is temporally at the eve of the Younger Dryas. During the period of the YD oak suffered a major decline in importance, going from 70% to 30% of the pollen assemblage, indicating a return to cool dry climatic conditions. *Pistacia* is still represented, perhaps suggesting that pistachio trees found a refuge in the Hula Valley, which, due to naturally high water tables, might not have suffered from severe droughts as much as other locations in the Levant during this time period. Wright and Thorpe (2003) point out that the grasses increase to 30%, but there is no corresponding increase in the chenopods and *Artemisia* as is found in other localities during the Younger Dryas. They interpret this as indicating less arid conditions in the Southern Levant than in more continental inland areas of the Near East and reconstruct it as an open oak savanna environment. The oak forest recovered at a time period dated to 10,372 BP (corrected and calibrated), seen at 1125 cm, although it never attained the abundance of the pre-YD forest.

Wright and Thorpe support this scenario by comparing the climatic implications of the Hula core with the oxygen isotope data from the Soreq
speleothems (Bar-Matthews et al. 1999) (see the discussion above in the section
"Oxygen Isotope Evidence"). The pronounced peak in $\delta^{18}O$ at around 19,000
BP represents the height of the Late Glacial Maximum and corresponds with the
date of the lower portion of the Hula core from 1560 to 1475 cm, in which the
pollen indicates similarly cool conditions. Just after this, from 18,000 to 13,200
BP, a major decline in the $\delta^{18}O$ curve points to the post-LGM warming that is
paralleled in the Hula core by the increase in forest from 1475 cm to 1240 cm
(at 17,400–12,900 cal. BP). The return to higher $\delta^{18}O$ levels in the Soreq record,
which marks the beginning of the Younger Dryas at circa 13,800–13,200 BP extending until circa 11,400 BP, marking the beginning of the Holocene, easily
corresponds with the major decline in oak pollen at Hula beginning after 12,900
cal. BP and ending sometime before 10,400 cal. BP.

Further support for the Wright and Thorpe interpretation of the Late
Pleistocene–Early Holocene phase of the Hula diagram comes from pollen
analyses at a number of Natufian sites from Israel and Jordan. Although there
is much room for critique of pollen studies from archaeological sediment, the
results of these studies reported by Leroi-Gourhan and Darmon (1991) have
an uncanny resemblance to the arboreal/nonarboreal sequences from Lake
Hula as redated in Wright and Thorpe (2003). Leroi-Gourhan and Darmon
report pollen results from the Pre-Natufian–Early Natufian transitional levels
at Hayonim Terrace, Israel (Level E) and Wadi Judayid (Bed J2c), Jordan. Radiocarbon dates on this level from Judayid (12,780 ± 660 bp) have large standard deviations and are calibrated to a long range of 17,000–13,500 cal. BP;
however, they generally fall into the lower portion of pollen zone 2 at Hula
during the transition from a dominance of chenopods and *Artemisia* to a
more forested environment. The pollen from Hayonim Terrace and Wadi Judayid both point to dry conditions dominated by chenopods. This might represent the very end of the LGM dry phase. This is followed by a humid phase
at the two sites, with a significant increase in oak and pine pollen. Hayonim
Terrace Level D occurs shortly after this initial amelioration and is dated to a
range of 14,400–13,500 cal. BP (11,920 ± 90 bp). It contains pollen of oak,
carob, and pistachio indicating a strong presence of woodland vegetation. The
dry-loving compositae go from 85% of the assemblage to 25% in this level.
Horowitz (1976) reports similar pollen evidence for a moist phase at the
Negev Natufian sites of Rosh Zin and Rosh Horesha with 9.5% arboreal
pollen in a locality in which no trees exist today. It represents a transitional

phase from a moist to dry climate. This site has two reliable radiocarbon dates of circa 12,800 cal. BP (10,880 ± 280 bp) and circa 12,300 cal. BP (10,490 ± 430 bp), which correspond very well to the terminal peak of forest pollen at Hula dated by Wright and Thorpe (2003) to 12,879 cal. BP at the transition between pollen zones 2 and 3, or the very beginning of the Younger Dryas dry phase. Pollen evidence for environmental deterioration associated with the Younger Dryas after this time period is present at Hayonim Terrace (Leroi-Gourhan and Darmon 1991) with diminishing tree pollen and the disappearance of aquatic species. Steppic plants dominate the spectrum. In Late Natufian levels of Salibya IX in the Jordan Valley, chenopods dominate, forming 90% of the assemblage, but drop down to 8% with the increase of arboreal pollen at the beginning of the Holocene PPNA levels (Darmon 1988, cited by Leroi-Gourhan and Darmon 1991).

Moving northward, Kuzucuoğlu and Roberts's (1998) summary of arboreal pollen curves for cores from Western Anatolia show post-LGM forest expansions similar to those of Hula at Karamik, Söğüt, Beyşhir II, Yeniçağa, and Ladik. These differ significantly from their summary of diagrams from Central and Eastern Anatolia (at Eski Acıgöl and Van), which indicate a dry steppe environment until the onset of the Holocene. This could be an indication of more continental climatic controls on these coring locations during the Late Glacial period. Rossignol-Strick's rephasing of Hula would place it more in sequence with these diagrams from central and eastern Turkey than with the more Mediterranean sequence of western Turkey. It seems more reasonable to accept the smaller corrections to the ^{14}C dates at Hula proposed by Cappers, Bottema, and Woldring (1998). This would establish the presence of *Pistacia* at an earlier date in the Hula Valley than in some other locations, perhaps as a result of a nearby refuge of the tree during the LGM and an earlier colonization of the Southern Levant.

Finally, if we accept Rossignol-Strick's (1995) and Wright and Thorpe's (2003) interpretations of the Ghab core, and Wright and Thorpe's (2003) and Cappers, Bottema, and Woldring's (1998) interpretations of the Hula core, the two cores become complimentary rather than out of phase. The main differences are that the Early Holocene forest at Ghab is at least as extensive as the pre-YD forest, but at Hula the Early Holocene forest never again attains the extent of the post-LGM forest. It also appears that the chenopods are less pronounced during the YD at Hula than during the LGM, and also less prominent than the YD at Ghab. *Pistacia* at Hula also appears at an earlier date than at Ghab. The milder YD at Hula compares favorably with the isotope data from the Soreq Caves, which indicate that the cool dry conditions of the LGM are far more pronounced than during the Younger Dryas. These variable factors may be related to local topographic and landscape trends in the Southern versus the Northern Levant. In the case of Lake Hula, the lack of full recovery of the forest from the impact of the Younger Dryas event might also be related to extensive burning or forest management on the part of Early Holocene populations (Roberts 2002; Wright and Thorpe 2003).

GEOMORPHOLOGICAL EVIDENCE
FOR PALEOLANDSCAPES AND HYDROLOGY

Evidence from isotope and pollen data suggests that shortly after the Pleniglacial cold dry episode in the Southern Levant there was a rapid warming with a significant rise in precipitation and a rapid expansion of the Mediterranean forest vegetation. The Younger Dryas event brought a return to cool dry conditions, although not equaling the magnitude of the LGM, but enough to cause the forests to contract again. With a return to a warm wet environment at the beginning of the Holocene, the Mediterranean forest recovered, but not to the same extent as the forest expansion after the LGM. These fluctuations in precipitation and vegetation had a profound effect on hydrological systems and landscape stability. Such shifts would leave a mark on the sediment record of the Levant, especially in marginal or ecotonal areas. In this section I review some of the geomorphological evidence for landscape and hydrological changes in this period.

In the Southern Levant there are two main stream terrace sequences relevant to the Terminal Pleistocene (see Goldberg 1986, 1994; Goodfriend 1987; Goodfriend and Magaritz 1988; and A. Rosen 1986b). In central Israel the oldest of these is designated the *Magalit Terrace* (A. Rosen 1986b). This terrace was dated to circa 23,500 cal. BP (23,300 ± 750 bp) in its upper levels by radiocarbon on $CaCO_3$ nodules within a paleosol (Goodfriend, personal communication, 1988) and represents the final stages of an alluvial floodplain directly preceding the Pleniglacial dry phase. At the type section along Nahal Shiqma in central Israel this terrace is about 10 m high and consists of alternating fine and coarse-grained alluvium indicative of fluctuating stream flow in a meandering stream system. Episodes of landscape stability are marked by well-developed paleosols characterized by high clay content, reddish color (5 YR and 7.5 YR), and large $CaCO_3$ nodules (see figure 4.5). The

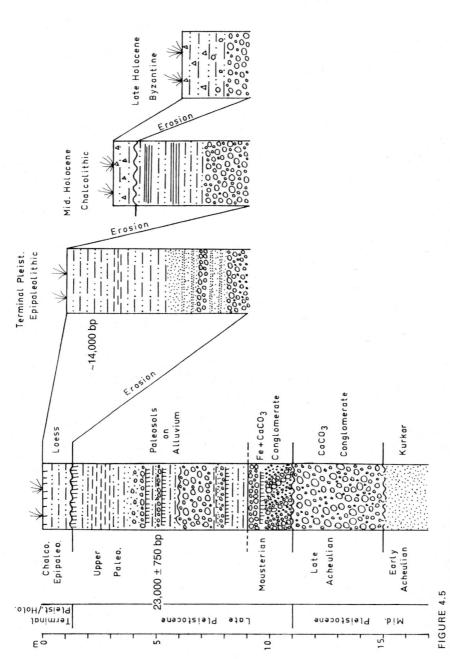

FIGURE 4.5
Section Drawing of Magalit Terrace (A. Rosen 1986b).

Magalit Terrace system is paralleled by terraces of similar age in the Negev Desert. Goldberg (1986, 1994) described Pleistocene alluvial terraces, which he dated by artifactual remains to the Upper Paleolithic period. These terraces are characterized by massive alluvial sand, clay, and silt interspersed with signs of standing water. The deposits represent actively alluviating floodplains in a low-energy stream regime that probably drained a dry steppe environment with water tables higher than those of today, a more steady winter rainfall, and possible summer rains. Goodfriend and Magaritz (1988) also dated $CaCO_3$ nodules from paleosols on alluvial terraces in the Negev Desert and identified floodplain soil horizons dating to circa 37,000 bp, 28,000 bp, and circa 15,500 cal. BP.

Sometime between 22,000 and 18,000 cal. BP, in both central and southern Israel, streams began to incise their channels, and the old floodplains were abandoned. This time period represents the cool dry Late Glacial Maximum. The decrease in rainfall and resulting drop in water tables led to decreased stream base flow and subsequent downcutting of the channels. The retreat of forest vegetation allowed for more erosive runoff and the stripping of soils with less water infiltrating the sediment and replenishing the water table.

Renewed alluviation resulted in a subsequent set of terraces in the Negev and Sinai associated with Epipaleolithic period sites dated by ^{14}C to circa 17,500–15,000 cal. BP (Goldberg 1994; A. Rosen 1986b). Goldberg describes these terraces in Kadesh Barnea as characterized by gleyed sands and silts. This is equivalent to the terrace system in central Israel described as the *Shiqma Terrace* (Rosen 1986a). The Shiqma Terrace system was widespread throughout the Shephela and inner coastal plain of central Israel and is dated by a radiocarbon determination to circa 15,500 cal. BP (Goodfriend, personal communication, 1988). It is characteristically about 8–10 m high and like the Negev terraces alternates between gravel deposits, laminated sands, sandy silts, and gleyed sediments (A. Rosen 1986b). The Shiqma Terrace is coarser grained than the Magalit Terrace, and the sediment color is more yellowish brown rather than red. It commonly contains the bones of Pleistocene fauna such as *Bos primigenius* (wild cattle) and *Equus hemionis* (onager), suggesting an open parkland or steppic environment. Stone tools with Epipaleolithic affinities were also found in numerous places in the sediments of this terrace. The alluvial system is characterized by aggrading floodplains, shifting channels, and complexes of levees and backswamps. Sometime after circa 14,000 cal. BP the stream regime shifted from one of widespread alluviation to extensive downcutting. The channels became narrow and deeply incised with no overbank flooding. This phenomenon was most likely a result of the drying associated with the onset of the Younger Dryas event and accompanying drop in water table.

The Negev Epipaleolithic terraces and Shiqma Terrace system in central Israel are indicative of the increasingly warm and wet climatic conditions that occurred after the very cold dry episode of the LGM. When we put this together with the isotope and pollen data, we can draw a picture of the Late Pleistocene landscape and environment that includes expanding forests and ultimately well-vegetated and forested hillslopes with open grassy parkland on the coastal plain and a more arid grassland environment in what is today the Negev Desert. This landscape was populated by herds of bovids and equuids as well as bands of hunter-gatherers. At the beginning of the Younger Dryas the forest rapidly declined with grassland areas expanding in the north and desertic conditions taking hold in the south of the country. Dropping water tables under drier climatic conditions led to the incision of stream systems from northern Israel down to the Negev Desert.

LACUSTRINE EVIDENCE
The most intensively studied lake in the Southern Levant is the Dead Sea, located today at about 415 m below mean sea level (mbsl), within the Jordan Rift Valley. Its northernmost shore is located about 10 km south of Jericho, in a harshly arid environment receiving an annual precipitation of 50–100 mm per year. The Dead Sea is a hypersaline terminal lake that takes in water from the Jordan River but has no outlet. It is a remnant of the much larger Pleistocene Lake Lisan. Pioneering work on Lake Lisan deposits by Neev and Emery (1967) and later Begin, Ehrlich, and Nathan (1974, 1980) showed that the lake once stretched from what is today Lake Kinneret (Sea of Galilee) down to the Hatzeva Springs in the modern Arava Valley and dated from circa 70,000–60,000 BP until circa 11,000 BP. Mid-nineteenth-century geologists believed Lake Lisan represented moister climatic conditions, and this was among the evidence cited to support the concept of low-latitude Pleistocene *pluvial* periods that geologists once thought accompanied the expansion of glaciers during cold episodes in Europe (Lartet 1865:798, cited by Flint 1971:19). Lake Lisan reached its maximum extent at an elevation of 160 m below sea level at circa 25,000 cal. BP. The lake then decreased in size by stages until around 11,000 cal. BP and was subsequently replaced by the much

smaller Dead Sea (Neev and Emery 1967; Bartov et al. 2002). The Lisan Lake deposits consist of the *Laminated Member* at the base, which is composed of bedded clays and sands. This is overlain by the *White Cliff Member* formed primarily of aragonite with lenses of diatomite and gypsum. Begin, Ehrlich, and Nathan (1974, 1980) conducted an extensive study of these deposits from a series of columnar sediment sections along the length of the Lisan formation. They demonstrated an increase in chemical sediment over detrital sediments from the lower Laminated Member to the upper White Cliff Member. This indicates steadily drying climatic conditions throughout the Late Pleistocene. Begin, Ehrlich, and Nathan (1974) also defined a later post-Lisan member, the *Unnamed Clastic Unit*, consisting of reddish sands, silts, and clays, which they dated between circa 18,000 and 13,000 cal. BP. They attribute this member to a renewed rainy period with greater alluvial runoff from circa 22,000 to 14,000 cal. BP, suggesting a renewal of moist climatic conditions, although not as rainy as the episodes that initially filled Lake Lisan. Subsequent to the Unnamed Clastic Unit, there was a long phase of drying in the lake basin (Begin et al. 1985).

More recently, work by Bartov et al. (2002) and Bartov et al. (2003), has enhanced the resolution of the climatic interpretation of Pleistocene Lake Lisan sediments. These researchers examined the exposed facies from alluvial, beach, delta, and off-shore sedimentary environments of the Lake Lisan deposits to record the intervals of high water represented by laminated aragonite and detrital silts versus coarser clastic deposits indicating the presence of ancient shorelines and low lake levels (see figure 4.6). They concluded that the lake reached its maximum level of about 164 mbsl between circa 26,000 and 23,000 cal. BP. Although there were high lake levels during the peak of the Late Glacial Maximum, these were indicative of a cooler climate and subsequent reduction in evaporation rates rather than moister climatic conditions. The lake level then began to drop, reaching about 300 mbsl at 15,000 cal. BP with a slight rise at around 17,000 cal. BP (Bartov et al. 2002; Bartov et al. 2003). The cold dry Younger Dryas was represented by a significant drop in the level of Lisan Lake. The authors suggest that the low Lisan level at this time was a function of the *Heinrich events* marking the rapid release of cold water into the North Atlantic due to the melting of glacial ice. These events maintained cold water on the surface of the Atlantic, which then spilled into the Mediterranean. The ultimate effect was an interruption of Mediterranean storm systems through a reduction in the strength and frequency of storms and

FIGURE 4.6
Jordan Valley with the Dead Sea and Shorelines of
Pleistocene Lake Lisan (After Bartov et al. 2003).

increasing evaporation rates (Bartov et al. 2003). Neev and Emery (1967) also attest to this very dry episode in the YD. They recorded a deposit of clay mixed with large halite crystals indicating a low lake level at 426 mbsl. The level yielded a ^{14}C date on wood calibrated to circa 13,600–13,100 BP (11,315 ± 80 bp). This unit was overlain by a deposit of 7 m of halite, which Neev and Emery interpreted as a phase of strong evaporation and low lake levels throughout the YD. Subsequent to this drying the lake refilled at the beginning of the Holocene, becoming the modern Dead Sea.

Goldberg (1994) summarizes evidence based on the location of prehistoric sites along the lake margins (see Hovers and Bar-Yosef 1987; Bar-Yosef 1987; Hovers 1989), the geoarchaeological sequence at Wadis Fazael and Salibiya on the western margin of the Jordan Valley, as well as associated pollen studies (Bar Yosef, Goldberg, and Leveson 1974; Schuldenrein and Goldberg 1981; Leroi-Gourhan and Darmon 1991). Prehistoric sites dating from circa 20,500–16,500 cal. BP, containing Kebaran and Geometric Kebaran industries, are located close to former lake shores between 180 mbsl and 203 mbsl in elevation, suggesting that Lake Lisan was close to its highest level until around 16,000 cal. BP. This concurs with Druckman, Magaritz, and Sneh (1987), who dated oolites and other organic remains by ^{14}C and also conclude that there was a high lake stand of 180 mbsl at circa 17,500 cal. BP (14,600 ± 240 bp).

The geoarchaeological sequences at Wadis Fazael and Salibiya show that Natufian sites are located within gullies and depressions in the eroded Lisan marl deposits at about 215 to 230 mbsl, indicating a dry phase and lowering of the lake level during the Natufian period (Goldberg 1994). This concurs with pollen taken from these sites that also indicates a wetter environment during the Geometric Kebaran and Early Natufian occupations of the valley, with increasingly arid conditions in Late Natufian times (Leroi-Gourhan and Darmon 1991). The environmental reconstructions of Begin et al. (1985), Bartov et al. (2002; 2003), Schuldenrein and Goldberg (1981), Goldberg (1994), and Leroi-Gouhran and Darmon (1991) all complement the Soreq isotope data in a scenario of moist conditions following the LGM, with a dry snap during the Younger Dryas event.

SUMMARY OF LATE PLEISTOCENE ENVIRONMENTAL SETTINGS

This review of recent data from oxygen isotopes, pollen, and geomorphology allows us to pull together a summary scenario of environmental conditions on the eve of the Holocene (see figure 4.7). Beginning with the end of the Late

Approximate Date cal. BP	Oxygen Isotopes	Carbon Isotopes	Pollen Northern Levant (Ghab)	Pollen Southern Levant (Hula)	Stream Activity	Paleosols	Lake Lisan Levels
10,000	Warm - Moist		Forest Readvance	Forest - Parkland	Valley Stabilization		- - - - -
11,000	Early Holocene			- - - - -	- - - - -		
				Dry			Lake Level
12,000	Cold- Dry		Dry Chenopod - Artemisia Steppe	Parkland Steppe	Erosion and Valley Incision		Dropping
13,000	Younger Dryas		- - - - -	- - - - -	- - - - -	- - - - -	Steadily
14,000	- - - - -	- - - - -			High Water Tables &	Paleosol	
	Warmer - Moister	Warmer - Moister (C₃ plants)		Forest	Valley Alluviation		
15,000				Expansion	- - - - -	- - - - -	- 300 mbsl
16,000			Forest				
	- - - - -	- - - - -	Expansion		Erosion and Valley Incision		- - - - -
17,000							Small Rise
				- - - - -			- - - - -
18,000	Cold - Dry	Cold - Dry					
19,000		(C₄ plants)		Chenopod - Artemisia Steppe			- 270 mbsl
20,000			- - - - -		- - - - -	- - - - -	- 220 mbsl
			Chenopod - Artemisia Steppe		Valley Alluviation	Paleosol	- 164 mbsl

FIGURE 4.7
Summary of Terminal Pleistocene Climatic and Environmental Changes in the Levant (Prepared by A. Rosen).

Glacial Maximum at around 18,000 cal. BP, climatic conditions in both the Northern and Southern Levant were generally very dry and cool. The cool climate reduced the evaporation rate, thus allowing the expansion of Lake Lisan in the Jordan Valley in spite of these very dry conditions. The Mediterranean zones of the Levant were dominated by steppic vegetation including chenopods and *Artemisia*, which are dry-land shrubs, as well as grasses indicating somewhat moister steppic conditions. Some trees such as oak, pistachio, and carob managed to survive in small stands or refuges on better-watered ground, possibly around springs or marshes, which existed due to the low evaporation rates. Streams that had been actively alluviating and building floodplains throughout this area before the LGM incised their beds as base flow decreased with the decrease in rainfall amounts. This was especially the case in the drier Southern Levant.

Rainfall and temperature began to steadily increase from circa 17,000 until circa 14,000 cal. BP. Forests began to progressively expand in the Mediterranean zones of both the Northern and Southern Levant. Grasslands increased in extent in the drier regions of the Southern Levant. Lake Lisan continued to drop in level due to higher evaporation rates, after a brief rise around 17,000 cal. BP. Streams in the Southern Levant increased in flow, and there were renewed phases of active floodplain buildup, including stable episodes of soil formation under moister steppic vegetation in the south of the region. This area also supported herds of wild equuids and *Bos primigenius*.

Subsequent to this clement phase of climatic amelioration, the Younger Dryas event abruptly took hold and reversed the trend toward warmer and moister climatic conditions with a renewal of cold dry climate. Although the Younger Dryas is clearly recognizable in many of the lines of proxy evidence available in the Near East, the climatic effect in the Southern Levant appears to be less severe than in locations further to the north. The Soreq Cave isotopic data show a clear reversal to cooler and drier conditions, although these appear to be not as harsh as those of the LGM. Nevertheless, the forests of the Northern Levant sharply declined, being replaced by dry steppic vegetation until the abrupt shift back to a forested landscape with the beginning of the Holocene. In the Southern Levant, according to the dates of Cappers, Bottema, and Woldring (1998) for the Hula pollen diagram, the forest retreated, although not as abruptly as indicated in the Ghab diagram. It was replaced by a parkland steppic environment. Streams in the Southern Levant began to reincise their beds, leaving the former lush floodplains abandoned and dry. Paleosol formation ceased under the drier and colder climate. The level of Lake Lisan continued to drop until the lake itself lost its identity as a large Pleistocene lake, becoming what we recognize today as the modern Dead Sea.

This picture of the Younger Dryas event landscape and vegetation abruptly gave way to the warming and wetting trend that accompanied the Holocene of the past 11,500 years. The Holocene defined the climatic parameters, created the obstacles, and offered environmental opportunities for some of the major events of human cultural development that led to our current modern societies. Before I turn to the way different cultural groups responded to these environmental changes, I consider the evidence for climatic variability throughout the Holocene in the Near Eastern Levant in the following chapter.

5

Holocene Paleoenvironments of the Near East

In comparing the evidence for climatic change between the Holocene and Pleistocene records, one might be forgiven for assuming that there has been no "significant" environmental change since the beginning of the Holocene in the Near East. Chapter 1 of this volume contrasts the meaning of *significant climatic change* as used by some paleoclimatologists with its usage by anthropologists and geographers studying ancient farming communities. As previously emphasized, a relatively small shift in rainfall patterns in a marginal farming community may make the difference between arable and nonarable land, or between high and low crop yields, at different levels of technological development. In the light of this, there have been many culturally significant changes in climate throughout the course of the Holocene. Although these may simply be represented on a curve of proxy data as small fluctuations, in fact, they could be highly significant for human societies.

Since we are concerned here with the possible impact of these changes on ancient communities, this chapter reviews the evidence for climatic fluctuations and environmental change during the Early (9500–5500 cal. BC), Middle (5500–2000 cal. BC), and Late (2000 cal. BC–present) Holocene, pointing out strong trends that may have presented stumbling blocks or opportunities to social groups in the region during different phases of the period. In the previous chapter, dates are reported in calendar years BP wherever possible. In this chapter on Holocene environmental change dates are calibrated in BC/AD since most archaeologists working in this time frame will find this more familiar.

EARLY HOLOCENE

The Early Holocene is defined here as the period from circa 9500 to 5500 cal. BC. It is a momentous period for the story of human cultural development. This is the time period that witnessed the very first agricultural villages worldwide, the emergence of cultural complexity, and the beginnings of a stepwise increase in the ability of humankind to make a major and lasting impact on the environment. Before exploring the role of the environment and climate change in these developments, it is necessary to outline the proxy evidence for the climate at the dawn of the Holocene and to propose a scenario for environmental change at this time.

Isotope Evidence

Goodfriend (1999) summarized isotope evidence for Holocene climatic change in the Levant. In his work on carbon isotopes he examined snail shells from the Negev Desert to determine if the snails were consuming vegetation typical of temperate or dry environments (Goodfriend 1988). The vegetation in temperate habitats is characterized by plants that use C_3 photosynthetic pathways to fix carbon, while the plants in arid zones primarily use C_4 pathways. Organic carbon becomes incorporated into the calcium carbonate of the snail shells when they ingest the plants in their surroundings. Calculation of the $\delta^{13}C$ from the snail shells provides values typical of either C_3 or C_4 plants.

The Negev Desert of the Southern Levant was selected for the use of this technique because of its marginality between temperate and arid zones. The region is therefore sensitive to minor shifts in rainfall amounts and a good location for tracking environmental change. Goodfriend found that in the Early Holocene, the snails were dining on C_3 plants in areas that are today typified by C_4 type plants. This indicates that throughout the Early and Middle Holocene the region received over 290 mm of rainfall per annum, whereas today the rainfall is well under 200 mm (see figure 5.1).

This evidence was reinforced by Goodfriend's work on the oxygen isotopes from snail shells in the same region. The $\delta^{18}O$ of the shell carbonate is closely related to the value for the precipitation at the time the shells were formed (Goodfriend and Magaritz 1987). The curve for $\delta^{18}O$ change throughout the Holocene is interpreted by Goodfriend as representing the oxygen isotope variations in Holocene precipitation (Goodfriend 1991). His results indicate that the earliest Holocene began with $\delta^{18}O$ values similar to those of the present.

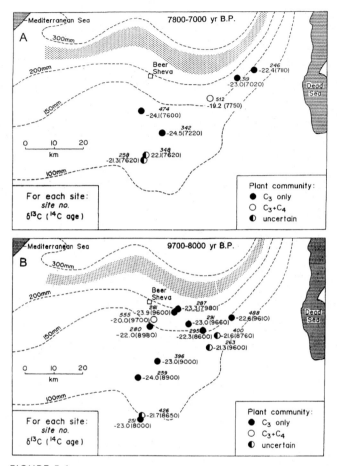

FIGURE 5.1

Early Holocene Distribution of Pure C_3 and Mixed C_3-C_4 Plant Communities in the Northern Negev Desert as Determined from $\delta^{13}C$ Values in Land Snail Shells. The shaded zone indicates the modern transition zone between these two types of plants. (A) Samples radiocarbon-dated to 7800–7000 BP. (B) Samples radiocarbon-dated to 9100–7000 cal. BC (From Goodfriend 1999).

From circa 8950 to 4800 cal. BC there was a steady decrease in values. Goodfriend attributes this to the increasing importance of a rainfall pattern originating in central Europe and sweeping southward across North Africa and finally into the Levant from the southwest, a pattern quite different from today's dominant rainfall cycle that originates in the North Atlantic and moves southeastward toward the Levant (Goodfriend 1999). Goodfriend (1991) interprets this early pattern as indicative of more frequent rainfall events and therefore a greater quantity of overall moisture in the Levant.

A more detailed record of Early Holocene climatic fluctuations comes from the work of Bar-Matthews, Ayalon, and Kaufman (1997) and Bar-Matthews et al. (1999) on $\delta^{18}O$ and $\delta^{13}C$ values from speleothems in the Soreq Caves of central Israel. Some of these results are discussed in the previous chapter in the review of evidence for Late Pleistocene climatic change. The climatic data for the Early Holocene show a clear drop in $\delta^{18}O$ values at the beginning of the Holocene from circa 9000 BC until circa 7500 BC (see figure 4.1 in the previous chapter). After this initial steady trend toward warmer and wetter climatic conditions there was a small reversal at circa 7000 BC and then a series of fluctuations throughout this Early Holocene (see figure 5.2). Two major episodes appear as very low $\delta^{18}O$ values at circa 6400 BC and circa 5600 BC. These values are the lowest that have been recorded over the course of the entire 60,000-year period examined by Bar-Matthews et al. (1999). The low points represent episodes of very warm and wet climatic conditions. The first of these at circa 6400 BC was extremely brief, probably lasting no more than a hundred years. The second period with its lowest point at 5600 BC could have lasted for a period of about 300–400 years. Rainfall estimates for this period suggest a yearly average ranging from 675 to 950 mm, almost twice the present-day rainfall averages (Bar-Matthews, Ayalon, and Kaufman 1997).

Between these very warm and wet Early Holocene episodes there is a small reversal peaking at around 6100 BC. This event is less evident in the $\delta^{18}O$ than in the $\delta^{13}C$ curve. It represents evidence for what is known in the paleoclimate literature as the Holocene *8.2KY event*. This climatic episode is strongly represented in the Greenland ice cores and is indicative of an abrupt cooling event that was manifested in North Africa and the Near East as a short but significant cool dry phase. According to the Bar-Matthews et al. (1999) data, a dry episode at circa 6100 BC separated the two moist periods at circa 6400 BC and 5600 BC.

Pollen Evidence

As discussed in chapter 4 of this volume, Rossignol-Strick (1995, 1999) reported on pollen studies from eighteen marine cores from the Mediterranean Sea and compared these with terrestrial records from Greece, Syria, Israel, Turkey, and Iran. In this study she identified two major horizons marking significant climatic changes in the Late Pleistocene and Early Holocene. These

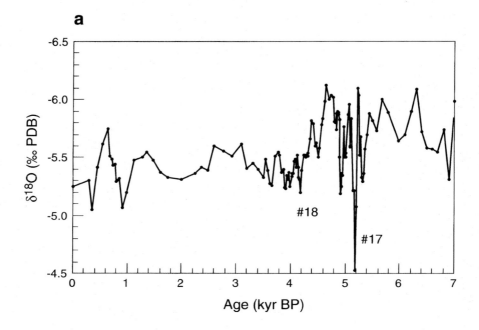

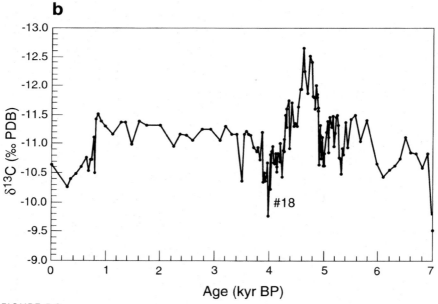

FIGURE 5.2

Graph Showing the δ¹⁸O and δ¹³C Values of Speleothems from Soreq Cave, Israel, Deposited during the Last 7000 Years. A sharp rise in δ¹⁸O values at 5200 BP is indicated as #17, and an increase in values from 4600 to 4000 BP indicated by #18 (After Bar-Matthews and Ayalon 2004).

are the Chenopodiaceae phase indicating the cold dry climatic conditions of the terminal Pleistocene Younger Dryas event, and the subsequent warm moist period of the Early Holocene, which she termed the *Pistacia* phase. In the marine cores this warm moist episode is identified by the formation of *sapropel* (sediments composed of organic residues). The dates for the earliest Holocene formation of the sapropel are given with a mean age of circa 7600 cal. BC (8548 ± 120 bp) with an oldest age of circa 8050 cal. BC (8960 ± 100 bp). The youngest ages of the sapropel range from circa 6600 cal. BC (7650 ± 140 bp) to circa 4900 cal. BC (5970 ± 170 bp).

The presence of *Pistacia* and increase in *Quercus* during this period in the marine and terrestrial cores is interpreted as indicating precipitation levels from 800 to 1300 mm per year, including regular summer rains, in contrast to today's situation of summer drought (Rossignol-Strick 1999:526, 528). The brief interval of the 8.2KY cooling event recognized in ice cores at 8200 BP (or ca. 6200 cal. BC) (Alley et al. 1997) and by Bar Matthews et al. (1999) from their $\delta^{13}C$ curve in speleothems at the Soreq Caves also appears in the pollen record. Rossignol-Strick points out similar indications of a cool episode from an abrupt but short-lived decrease in *Quercus*, *Pistacia*, and other deciduous trees at circa 6900–6400 cal. BC in the land record, and a 200-year cessation of sapropel development in the Adriatic Sea (Rohling, Jorissen, and De Stigter 1997; Rossignol-Strick 1999).

These results from the Mediterranean Sea and terrestrial littoral region seem to contrast with key pollen diagrams from eastern Turkey (Lake Van) and western Iran (Lake Zeribar). In the case of Lake Van, Rossignol-Strick suggests that in the van Zeist and Woldring (1978) diagram (a combination of two cores), the Chenopodiaceae and *Pistacia* phases are displaced. Her interpretation can be supported by the $\delta^{18}O$ record at Van (Lemcke and Sturm 1997; Wick, Lemcke, and Sturm 2003), which indicates a marked dry episode during the Younger Dryas between 12,600 and 10,400 BP (varve chronology) with a rapid Early Holocene recovery to moist climatic conditions.

Kuzucuoğlu and Roberts's (1998) summary of pollen diagrams from Turkey were redrawn to illustrate the arboreal/nonarboreal pollen curves. The Early Holocene pattern very distinctly shows a rapid and copious recovery of arboreal species just after the termination of the Younger Dryas. This is especially true for the westernmost core sites. The eastern sites including Eski Acıgöl and Lake Van demonstrate a slower recovery of the forest, even given the dating correction of Rossignol-Strick. This difference in the onset of the

Holocene forest may be a function of differing climatic patterns and rainfall regimes, with eastern Turkey, western Iran, and localities with higher altitudes more strongly influenced by continental conditions than the sites in western Turkey and the Levant, which fall squarely within the Mediterranean climatic system.

Pollen analyses for the Terminal Pleistocene to Holocene at Lake Hula were first published by Horowitz (1971). Since that time several different researchers have conducted pollen analyses from this important catchment basin. M. Tsukada's diagram (published in English by van Zeist and Bottema 1982) was the most detailed schematic summary of Late Pleistocene and Holocene pollen until a more recent core published by Baruch and Bottema (1991, 1999) (see figure. 4.4 in the previous chapter). As mentioned in chapter 4 of this volume, the arboreal pollen curve from the portion of this core covering the Late Pleistocene agrees well with the $\delta^{18}O$ data from the Soreq Caves.

At the beginning of the Holocene (attributed to upper pollen zone 3 by Wright and Thorpe 2003), the curve for arboreal pollen remains relatively low (fluctuating between about 35% and 50% as opposed to 60% before the Younger Dryas), with many minor fluctuations. Tree pollen stays at this moderate level until circa 5500 cal. BC, after which it dips to an even lower level at the beginning of the Middle Holocene in mid–zone 5. The Early Holocene in the Hula basin is generally characterized by a mix of trees and shrubs with a slight rise in evergreen oaks (*Quercus calliprinos*) and an early rise then decrease in deciduous oak (*Quercus ithaburensis*). *Pistacia* remains a significant component of the forest. Grass and cereal pollen also display minor variations but continue to be relatively high throughout the Early Holocene.

The Niklewski and van Zeist (1970) diagram for Ghab with corrected chronology shows a rapid and complete recovery for the forest immediately after the cessation of the Younger Dryas event. A major forest expansion, including a significant component of *Pistacia* (zones Z_1 and Z_2) occurring in the Early Holocene indicates rapid warming and wetting of the post-Pleistocene environment.

Yasuda, Kitagawa, and Nakagawa's (2000) Ghab diagram from northwestern Syria best correlates with the Hula diagram if we assume that the radiocarbon dates need correction for the *hard water error* (see figure 4.3 in the previous chapter) (Cappers et al. 2002; Wright and Thorpe 2003). According to Wright and Thorpe (2003), the pollen profiles representing the beginning of the Holocene are located in the upper portion of Yasuda, Kitagawa, and Nakagawa's zone 1 at a depth of approximately 550 cm. This shows a rapid influx of oak, pine, and pistachio, again indicating warming and wetting. Oak reaches its peak levels throughout zone 2, then abruptly drops.

The pollen profile at Lake Hula shows a modest increase in arboreal pollen at the beginning of the Holocene, while the Ghab profile from both cores indicates a much more significant rise. One possible explanation for the low level of forest expansion in the Early Holocene at Hula could be the influence of humans on the landscape. At Hula in the first portion of zone 4, the forest began to reexpand after the decline in the Younger Dryas event of zone 3. This reexpansion came to an abrupt halt in mid–zone 4. It is possible that human inhabitants of the region locally controlled forest growth to increase open land for the growth of wild and cultivated cereals, as well as for the propagation of wild game and later for grazing land. This possibility is especially likely considering the $\delta^{18}O$ evidence from the Soreq Caves that points to this Early Holocene phase as being the warmest and wettest climatic episode of the entire Holocene, and, therefore, one would expect forest expansion under natural conditions. At Ghab, deciduous oak suddenly declined during the latter portion of the Early Holocene in pollen zone 3, with an increase in evergreen oak. This coincides with a large increase in microcharcoal at the same level and an increase in olive. Both of these are a strong argument for human interference in the natural vegetation patterns, resulting in a conversion of the oak forest into maquis (Yasuda, Kitagawa, and Nakagawa 2000).

Lacustrine Evidence

Paleoclimatic research has shown that lakes in desertic and semiarid localities in North Africa and the Arabian peninsula significantly expanded during the Early Holocene from circa 8000 to 4000 BC (Gasse 2000; Street-Perrott and Perrott 1993; Scott 2003). This supports the general picture of a dry terminal Pleistocene and a strong reversal to moister climatic conditions during the Early Holocene.

Research on the Dead Sea also indicates a significant recovery of the lake after the severe desiccation associated with the Younger Dryas and the near disappearance of the Pleistocene Lake Lisan (see chapter 4). Neev and Emery (1967) recorded 7 m of halite deposited in a core from the Dead Sea that dated back to the Younger Dryas, post-15,000 cal. BC. This would put the level even lower than that of the present-day Dead Sea level, which led Neev and Emery to suggest that the lake itself had completely dried. Other researchers agree that the lake was much lower at the end of the Pleistocene but did not completely disappear (Begin et al. 1985). Subsequent to this the lake level began to rise again. Neev and Emery (1967) put the lake level at close to that of the modern level (410 mbsl) at 10,000–8700 BC (9850 ± 150 bp). The Early Holocene deposits are described as gray lake clay and indicate a return to higher lake levels. These are dated by ^{14}C on wood to 7610–7290 BC (8440 ± 95 bp) and 7480–7080 BC (8255 ± 70 bp). Yechieli et al. (1993) examined sediments from a borehole along the modern shore of the Dead Sea with similar results. Their findings indicated an unconformity (a break in the sediment sequence indicating an erosional episode) between the massive Late Pleistocene salt layer and the subsequent return to deposition of clay-rich sediment indicating moist conditions and a higher lake level. These lake clays date to 7610–7290 BC (8440 ± 95 bp). High Holocene lake levels existed in the Dead Sea until the Middle Holocene.

Geomorphological Evidence

The geomorphological evidence for climatic and environmental conditions at the beginning of the Holocene is rare in the Southern Levant. This is due partly to the fact that much of the Early Holocene deposits were subsequently eroded from hillslopes and stream banks, and/or buried by later deposits in the valley bottoms. On the coastal plain of Israel Gvirtzman and Wieder (2001) investigated Early Holocene red Mediterranean soils known locally as *Hamra soils*. These are indicative of warm wet environmental conditions in the Mediterranean zone. One of these paleosols dates to approximately 8,000–5,500 cal. BC and represents paleosol development on the coast during the Early Holocene warm moist episode. Following this period, there was a renewal of windblown sediment deposits, suggesting drier environmental conditions.

Other geomorphological evidence comes from isolated pockets of sediment, often investigated in association with geoarchaeological studies of specific archaeological sites. In northern Israel, geoarchaeological studies at the Neolithic sites of Tel Teo in the Hula Valley and Tel Yosef in the Harod Valley identified soil horizons in marshy floodplain sediments dating from the Pre-Pottery Neolithic B-C period (ca. 6600–6250 cal. BC), overlain by colluvial deposits from the Pottery Neolithic (ca. 5800 cal. BC). The Early Holocene deposits indicate high water tables in a landscape inferred to have been associated with forested hillslopes and moister climatic conditions. The overlaying sediments suggest a reversion to more colluvial runoff in the sixth millennium BC (A. Rosen 2001, in press).

Schuldenrein and Goldberg (1981) reported on Neolithic period sediments at the sites of Salibiya and Fazael in the Jordan Valley. At some time in the Early Holocene, there was a rapid aggradation of *pink silts.* The source of this deposit was a mixture of wadi alluvium and hillslope wash or colluvium. Blocks of tufa spring deposits indicate that abundant water flowed in the Jordan Valley contributing to high water tables and the formation of marshes that existed sporadically during Natufian times but were more abundant during the Pre-Pottery Neolithic. Intensification of moist conditions in the Early Holocene led to increasing sedimentation in the later phases of the PPNA site of Salibiya IX and immediately postdating the site. The deposition also postdates the PPNA site of Netiv Hagdud (Goldberg 1994).

Goodfriend dated sediment sections from a number of localities in the Negev Desert. His results indicate that the Early Holocene deposits are primarily colluvial rather than fluvial in origin. He attributes this pattern to increased rainfall at this time, which encouraged colonization by burrowing animals. These animals loosened the deep Pleistocene loess deposits, which were transported down the hillslopes by the increased rainfall (Goodfriend 1987). There is a notable absence of sediment deposits of any kind for the period of circa 6000–7000 BC. Goodfriend attributes this to a very arid phase with little rainfall. Such an arid phase might have been associated with the now well-known cold dry 8.2KY BP event.

In Jordan, Cordova (2000) examined sediments from the dry steppic region of the central plateau. He described Early Holocene alluvial deposits from the Wadi ath-Thamad and Wadi Ar-Riwaq. The upper terrace in these wadis is derived from colluvial deposits with a date in the top levels of 7055–6600 cal. BC (7910 ± 70 bp), contemporary with the PPNB. Hunt et al. (2004) define the Early Holocene Faynan Member in Wadi Faynan further south in the Wadi Faynan. This is an alluvial deposit of cross-bedded silts and sandy gravels typical of meandering stream depositional environments. Pollen from these deposits indicates the moister environmental setting of a steppic zone, just at the margins of the Mediterranean forest. Their pollen evidence also indicates an interval of drier conditions coinciding with the event circa 8000 BP. McLaren et al. (2004) identified the Early Holocene Faynan Member in the Wadi Dana of the Faynan area. They suggest that these deposits indicate low-magnitude perennial rivers with a steady flow and a stable floodplain environment. This indicates wetter conditions in the Early Holocene between circa 7500 and 6000 cal. BC. Subsequently, windblown deposits after circa 5400 cal. BC indicate aridification and/or human impact on the landscape.

Early Holocene Summary

The Early Holocene saw moister climatic conditions with rapid warming and wetting after the end of the Late Pleistocene Younger Dryas event. Geomorphological evidence for landscape stability in the form of soil development on the coastal plain and inland Mediterranean zones, as well as perennial streams in Jordan, suggests there may have been an even distribution of moisture throughout the course of the year, with overall increased rainfall including gentle winter and summer rains. This type of rainfall regime would have been conducive to the growth of vegetation, which in turn would have increased the infiltration of moisture into the groundwater tables, inhibiting erosive runoff, and thus providing for abundant springs, marshlands, and perennial flow in stream valleys. This proposed abundant but gentle rainfall regime reflects the possible influence of a northward extension of the monsoon belt, which impacted the Arabian Desert in the Early Holocene and led to the formation of lakes in Arabia and North Africa (Gasse 2000; Goldberg and Rosen 1987; Hunt et al. 2004; McClure 1976; Street-Perrott and Perrott 1993; Scott 2003). In the Negev Desert, colluvial deposits also suggest moister conditions and increased activity of burrowing animals. The evidence for colluviation associated with Pottery Neolithic sites in the Mediterranean zone is most likely a combination of increased rainfall and some small-scale localized deforestation on the part of the Pottery Neolithic inhabitants.

MIDDLE HOLOCENE

The Middle Holocene is defined here as the period from 5500 to 2000 BC. It is a time that witnessed drier episodes after the Early Holocene moist phase, a return to moister climatic conditions, and then the beginning of the aridity that generally characterizes the climate of the Late Holocene up through modern times. In terms of human civilization the Middle Holocene gave rise to the first complex societies in the Near East. These societies were for the most part well adapted to their specific environments, some of which were marginal for farming activities.

Isotope Evidence

In examining the isotope curves from Middle and Late Holocene times in an area with a long and detailed archaeological record, there is a great temptation to accept the timescales at face value with no correction, and to link the record up with the much higher resolution of archaeological periodization. This of course has its obvious pitfalls since the correction of dates such as those from the Soreq Cave data might be in the neighborhood of hundreds of years. This makes it very difficult to assign climatic events to cultural actions, such as an acute dry spike with a known period of abandonment. Even worse, there is the temptation to "wiggle-match" the isotope curves with population and site settlement graphs. This is especially problematic in the Middle and Late Holocene when we are dealing with complex societies whose responses to climatic change are much more multifaceted than those of earlier societies, and the abandonments of settlements could be due to political rather than climatic factors (Rosen and Rosen 2001). It is far more productive to examine the curve in terms of phases of climatic stability versus rapid climatic fluctuations. Data from the Soreq Caves show that the Middle Holocene is a period with significantly greater variability in wet and dry phases than any time afterward (see figure 5.3). This fact alone provides us with important insights into the nature of the early complex societies that developed and flourished under these difficult and often unpredictable farming environments. This aspect of the social impact of climatic change is explored further in chapter 7. However, having warned the reader against the pitfalls of correlating specific climatic peaks and depressions with particular archaeological periods, I cautiously proceed to compare the two data sets as they are understood in their current respective timeframes with the proviso that future fine-tuning of the chronologies might offer slight shifts in the following tentative interpretations, and therefore, like bathing in the Dead Sea, the results must be accepted with a sizeable portion of salt.

The $\delta^{18}O$ isotope evidence from the Soreq Caves indicates a warm moist episode throughout the Early Holocene with optimal warmth and humidity near the end of this phase at circa 5500 BC (Bar-Matthews et al. 1999), roughly corresponding with the middle of the Pottery Neolithic period. Shortly after the beginning of the Middle Holocene, this climatic optimum had an abrupt reversal at circa 5000 BC leading to drier conditions than those of the present. A moister environment returned at a time approximately corresponding to the beginning of the Chalcolithic period at circa 4500 BC (Bar-Matthews and Ayalon 2004).

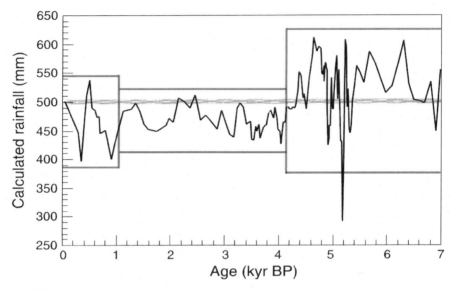

FIGURE 5.3
Paleo-Rainfall Amounts during the Last 7000 Years Calculated from Isotope Values from the Nahal Soreq Cave Speleothems (Figure After Bar-Matthews and Ayalon 2004). The values have been divided here into three phases of frequency and amplitude of rainfall fluctuations.

These moist conditions alternated with drier episodes through the Early Bronze Age (EB) I. There was a distinct rise in $\delta^{18}O$ indicating a very dry phase between circa 3100 and 3000 BC, approximately at the transition between EB I and EB II in the Israeli sequence. The return to low $\delta^{18}O$ values from circa 2900 to 2600 BC indicates a return to moister conditions late in the Early Bronze II. After this time period there was another acute dry episode around 2600 BC, followed by a steady return to very moist conditions from circa 2500 to 2200 BC (EB III). Finally, very dry conditions set in at the end of the Middle Holocene at circa 2200–2000 BC, corresponding to the end of the Early Bonze III period. The most important thing to note for both the Early and Middle Holocene episodes is the marked fluctuation in climatic conditions from dry to wet phases. These fluctuations are much less pronounced throughout most of the Late Holocene.

Although less detailed, Goodfriend's isotope data from snail shells in the Negev Desert (Goodfriend 1988, 1991, 1999) concur with this general scenario of wet environmental conditions during the third millennium Early Bronze Age period. His $\delta^{13}C$ determinations showed that from circa 5400 to

1200 cal. BC the southernmost distribution of C₃ (temperate-zone) plants was at least 20–30 km farther south than at the present time. This indicates a southward shift of rainfall isohyets and suggests that the Negev Desert received approximately 290 mm of rainfall in this area, which today receives only 150 mm per year. This represents almost double the annual average (Goodfriend 1988; Goodfriend and Magaritz 1987) (see figure 5.4). Goodfriend's oxygen isotope values from Negev land snail shells established that a minimum value of $\delta^{18}O$ occurred in the Middle Holocene at circa 5000 cal. BC. Goodfriend (1991, 1999) suggests that this indicates greater rainfall as well as a change in the source of the rainfall from the present-day pattern, in which Mediterranean rainfall originates in the North Atlantic, to a rainfall source in Eastern Europe. After this moist isotopic low there is a steady increase in $\delta^{18}O$ values until circa 3200 cal. BC, when modern values are attained and drier Late Holocene conditions set in.

Frumkin, Ford, and Schwarcz's (1999) work on the speleothems of caves in the Judean Hills also records a Middle Holocene drop in $\delta^{18}O$ values. Although these are not as precisely dated as the evidence from the Soreq Caves or the Negev land snails, an interpolated date would fix this drop at circa 2300

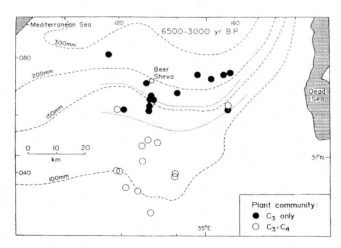

FIGURE 5.4
Middle Holocene Distribution of Pure C₃ and Mixed C₃-C₄ Plant Communities in the Northern Negev Desert from $\delta^{13}C$ Values of Organic Matter in Land Snail Shells. The two dotted lines correspond to the circa 240 and 290 mm mean annual rainfall isohyets in the middle holocene (From Goodfriend 1999).

cal. BC, with a subsequent sharp rise in values indicating increasingly drier conditions until modern values are obtained at circa 1300 cal. BC.

Pollen Evidence

The pollen diagrams from Hula (Baruch and Bottema 1999), Ghab (Niklewski and Van Zeist 1970; Yasuda, Kitagawa, and Nakagawa 2000), and Birkat Ram in the Golan Heights (Schwab et al. 2004) are the most complete and well-dated pollen records for the Middle Holocene of the Levant. In interpreting the Hula and Ghab diagrams this discussion uses the radiocarbon dates corrected for the hard water effect by Cappers, Bottema, and Woldring (1998) (Cappers et al. 2002; see also Wright and Thorpe 2003) and includes the calibration. The Hula diagram displays a marked increase in olive pollen beginning in the lower portion of zone 5 at around 6200 BC, corresponding to the Pottery Neolithic period (see figure 4.4 in the previous chapter). This is likely to represent the beginning of olive horticulture in this region. There is an expansion of the oak and pistachio components of the forest in zone 6 from circa 4500 cal. BC, extending throughout the Middle Holocene. Both deciduous and evergreen oak increase in this zone, with evergreen oak decreasing somewhat near the upper portion. *Pistacia* also expands at this time, perhaps testifying to warmer and wetter climatic conditions. The olive peak declined shortly after the beginning of this oak expansion at the end of zone 5. The oak declined around 1900 BC (extrapolated), perhaps corresponding to the shift from the many Middle Holocene moist phases to the beginning of the drier climatic regime in the Late Holocene. To support this, the pollen evidence shows a small but significant rise in *Artemisia* and Chenopodiaceae, perhaps also a result of warmer drier conditions.

Baruch's (1986) diagram of Lake Kinneret (Sea of Galilee) also indicates an expanded Middle Holocene forest with arboreal pollen values ranging from 45% to 60% at the bottom of subzone X_1, circa 3550–1600 BC (5220 ± 505 bp), calibrated and corrected for the hard water effect according to Thompson et al. (1985). Baruch suggests that the decline of olive at the top of subzone X_1 is associated with a marked decrease in olive production at the end of the third millennium BC, at a time when there was a general abandonment of EB III cities in the Southern Levant.

Pollen from Birkat Ram in the Golan Heights produced similar high pollen levels for deciduous oak (*Quercus ithaburensis*) (Schwab et al. 2004). After a sharp drop in levels at 4500 cal. BC, oak pollen remained at high concentra-

tions until a sharp drop due to deforestation in the Late Holocene Persian period after 586 BC.

In the Ghab diagram from northwest Syria by Niklewski and Van Zeist
(1970), the uppermost zone (Z_3) is interpolated by the authors to circa 4600 BC.
Even considering Rossignol-Strick's reevaluation of the dates at Ghab, this uppermost zone can be placed generally within the Middle Holocene, and the high
percentages of arboreal pollen concur with those of both Lake Hula and Lake
Kinneret in representing a picture of dense forests at this time. The more recent
Ghab core by Yasuda, Kitagawa, and Nakagawa (2000) has many more radiocarbon dates for the Middle Holocene (see figure 4.3 in the previous chapter).
If we correct these dates for the hard water effect according to Cappers et al.
(2002) by subtracting 1000 years, and calibrate the dates to calendar years, then
the portion of the diagram covering the Middle Holocene is zone 4 and the lowermost section of zone 5. These zones indicate high percentages of oak forest as
at Hula and Kinneret from circa 5600 cal. BC to 2000 cal. BC. Olive pollen has
its highest peak at the beginning of this period during the Pottery Neolithic and
then declines in the Chalcolithic period after circa 4400 cal. BC at both Hula and
Ghab. Evidence for deforestation during the Pottery Neolithic comes from the
decline in pine and cedar from higher elevations at Ghab, and the increase in
evergreen oak, indicating the spread of maquis (Yasuda, Kitagawa, and Nakagawa 2000). This may correspond to the sediment evidence for colluvial deposits during the PN in the Southern Levant. At the beginning of zone 5, dating
to the end of the Middle Holocene at circa 2000 cal. BC, there is a rapid increase
in marsh plants such as cattails (*Typha* sp.) and others. This is a strong indication of a drop in lake level and expansion of the marsh, most likely associated
with the climatic change to warm dry conditions that ushered in the Late
Holocene at this time (Yasuda, Kitagawa, and Nakagawa 2000).

Lacustrine Evidence

Neev and Emery (1967) conducted extensive analyses on the Holocene
portion of their Dead Sea cores. They found a sequence of lake clays interfingering with halite deposits, indicating periods of very low lake levels.
The lowermost portion of their Holocene core (Neev and Emery 1967;
Neev and Hall 1977) is in lake clay deposits and is dated to circa 9300 cal.
BC (9850 ± 150 bp). These lake clays represent the return of Lake Lisan as
the modern Dead Sea during the Early Holocene wet phase. This period of
high lake levels is followed by a very dry episode in which the lake level

drops significantly and halite is deposited. A return to high lake levels has one date of circa 3000 cal. BC (4410 ± 320 bp), which is Middle Holocene and archaeologically at the time of the Chalcolithic to Early Bronze Age occupation of the Southern Levant.

Frumkin and colleagues (Frumkin et al. 1994; Frumkin 1997) examined Middle Holocene lake level changes in the Dead Sea through a collection of rafted wood samples from the Mount Sedom caves. These samples date from approximately 5900 cal. BC and extend throughout the Middle and Late Holocene until circa 1700 AD. The absence of caves dating earlier than 5900 BC might be due to the submergence of the Sedom rock salt diapir where the caves are located (Frumkin 1997). The results of this study show an absence of rafted wood from circa 5800 to 4200 BC cal. Frumkin (1997) identifies this period as his "Stage 2," indicating low lake levels. A large number of samples cluster between circa 4200 and 2300 cal. BC, suggesting high lake stands in "Stage 3," with a maximum between circa 3500 and 2700 cal. BC. This evidence for a Middle Holocene high stand of the Dead Sea is supported by sediment data from the Nahal Ze'elim boreholes, which demonstrated submergence by the lake at this time, from the presence of clay, sand, and aragonite (Yechieli et al. 1993). The data from the Mount Sedom caves indicate that the Dead Sea reached its highest level in the last 7000 years during this Middle Holocene Stage 3 episode. Frumkin et al. (1994) suggest that the climatic amelioration began during the Chalcolithic period and lasted throughout most of the Early Bronze Age, coming to an end in the latter portion of the EB III period.

Klinger et al. (2003) combine geomorphological evidence with lake level data from the Dead Sea in their study of sediment deposits from alluvial fans in the Dead Sea basin. They found that alluvial terraces form during intervals of low lake levels and suggest that these represent the results of upstream erosion during the shift from wet to dry climatic episodes. Their results show that their "Q3b" terrace probably formed during the period of low lake levels between around 5000 and 4200 cal. BC, corresponding to Frumkin's (1997) Stage 2. Subsequently, the Stage 3 transgression is marked by beach bars on Q3b, indicating high lake levels at circa 2900 cal. BC (Klinger et al. 2003).

Geomorphological Evidence

Geomorphological evidence overwhelmingly suggests that stream alluviation increased during the Middle Holocene. Stream sedimentation can be a result of several different factors. These include climatic change, tectonics, human influence, and natural riverine cycles (Macklin, Lewin, and Woodward 1995). It is not often clear which factor was operative in bringing about changes in stream systems, and sometimes more than one factor is responsible. During the Pleistocene and Early Holocene there is little evidence that human influence played a major role in altering the alluvial regime of streams in the Levant. However, when we consider the Middle Holocene through modern times it becomes an important potential factor that must be discounted before geomorphological evidence can be used as a proxy for climate change. Today, human-induced erosion and valley sedimentation in semiarid regions often leave a signature on sediment sequences in the form of poorly sorted, coarse-grained sediment deposits. Therefore, in attempting to understand the cause of stream terraces in the Near East from Middle Holocene times, it is necessary to take into account the nature of the sediment lithologies and the spatial distribution of deposits throughout the stream system.

The Middle Holocene 'Erani Terrace is a system of low, 2–3 m high terraces located in the Shephela foothills and coastal plain of central Israel (A. Rosen 1986c, 1986b, 1991, 1997b). These terraces are situated within the previously eroded talweg of the Terminal Pleistocene channels. The deposits date to the Chalcolithic and Early Bronze Age by both radiocarbon determinations and pottery inclusions (A. Rosen 1991) and have a distinctive lithology. The sediment deposits consist of alluvial channel gravels in a lateral association with sets of 1 mm thick silt and fine sand laminations. These fine laminations are indicative of overbank levee deposits produced by streams with moderate competence that flow at a relatively steady rate. This depositional environment is very different from the present system of flash flooding, which deposits coarse-grained, very poorly sorted sands and gravels within the modern drainage channels. These Chalcolithic/EB I deposits can be interpreted as an increase in stream flow, but without the inclusion of coarse sediments that would be a consequence of soil stripping and erosive runoff. This enhanced stream activity was probably due to greater rainfall that was evenly distributed throughout the rainy season, rather than the heavy storms and flash floods that characterize today's rainfall regime. The 'Erani Terrace system is distributed from the Shephela foothills throughout the inner coastal plain of Israel.

Goldberg (1987, 1994) described a similar 2–3 m high Middle Holocene terrace system from the northern Negev near the Chalcolithic site of Shiqmim in the Nahal Beer Sheva wadi. This terrace is dated to the Chalcolithic period

(4200–3600 cal. BC) based on inclusions of sherds and the interfingering of wadi sediments with in situ occupation levels at the site. It consists of alluvial sands and silts that overlie more than 3 m of well-rounded gravels. When this terrace formation ceased to form, erosion and downcutting became the dominant regime (A. Rosen 2006). There is additional evidence for Chalcolithic/EB I period alluvial deposits in northern Israel near the site of Megiddo, dating to 3600–3400 cal. BC (4740 ± 45 bp) and 4200–3800 cal. BC (5200 ± 60 bp) (Rosen 2006). These Chalcolithic/Early Bronze Age terraces compliment other proxies for moister environmental conditions at this time.

The 'Erani Terrace and similar alluvial deposits from this time period have a significance that goes beyond environmental reconstruction. They also have important implications for the economy and subsistence agriculture of these archaeological periods. The terraces indicate that during each rainy season the streams of central and southern Israel overflowed their banks and deposited silt and fine sand across an active floodplain. These silts and moist floodplains provided an ideal opportunity for high-yield floodwater farming, which would have been more reliable than depending exclusively on rainfall farming to feed the growing towns and cities of this period. Phytolith evidence suggests that floodwater farming was indeed employed by peoples of these periods (A. Rosen 1995), a subject that is treated in more detail in chapter 7.

At the end of this phase of alluviation, streams throughout the Southern Levant began to incise their beds, creating deeper narrower channels and reducing the occurrences of overbank flow. This precluded the possibility of employing floodwater farming as either a principle or supplementary form of agriculture. The cause of this downcutting is most likely related to the climatic change around 2200 BC toward drier environmental conditions as indicated by the rise in $\delta^{18}O$ values from the Soreq Caves, the drop-off in deciduous oak pollen from the Ghab and Hula pollen diagrams, and the drop in the Dead Sea level at the close of the third millennium cal. BC (Bar-Matthews and Ayalon 2004; Baruch and Bottema 1999; Frumkin 1997; Yasuda, Kitagawa, and Nakagawa 2000).

This alluvial episode has contemporary parallels throughout the Near East. Extensive occurrences of Middle Holocene alluvial deposits were also described for southeastern Turkey in the fourth and third millennia BC, the upper Khabur region and Balikh Valley in northeastern Syria, and the Syrian Euphrates (Courty 1994; A. Rosen 1997c, 1998; Wilkinson 1998, 1999).

LATE HOLOCENE

The Late Holocene is defined here as the period from circa 2000 BC through the present. The part of the period before written records is probably the most challenging of all time stages for the reconstruction of changing climate. This is a direct result of the enormous human impact on natural ecosystems with increasingly intensive agricultural and pastoral economies. Land was cleared of forests, cultigens replaced naturally growing wild plants, soil erosion increased, and landscape features were altered. The result had a direct influence on proxy data from pollen cores and geomorphological evidence. It is therefore a challenge in paleoenvironmental reconstruction to sort out the cultural from natural factors that operate on the available proxy data. Isotope data from sea cores and speleothems are among the most reliable proxies since these are least affected by human activity.

On the other hand, this Late Holocene is probably one of the most important for understanding the sustaining of complex civilization with its associated population growth, emergence of empires, and changing human-land relationships including agricultural intensification, cash cropping, and the expansion of intensive agricultural communities into semiarid marginal regions. The role of climate was a key factor that had to be accounted for in the further development and expansion of complex societies.

Isotope Data

Goodfriend's $\delta^{18}O$ data from snail shells in the Negev Desert show a general warming and drying trend that is dated from circa 3000 BC to 1500 AD (Goodfriend 1991, 1999). However, the resolution of Goodfriend's data is too coarse to allow an accurate account of the smaller climatic fluctuations that may have impacted human societies in different archaeological periods. A more detailed record is again offered by the speleothems from Soreq Cave. According to Bar-Matthews' and Ayalon's (2004) graph of $\delta^{18}O$ determinations, the Late Holocene began with a sharp reversal from the moist episodes of the fourth and third millennia BC, to become what is essentially today's semiarid conditions in the Levant. This abrupt transition is the brief episode that gravely impacted the early state societies of the Near East at the end of the Early Bronze Age at the close of the third millennium BC (H. Weiss et al. 1993; A. Rosen 1989, 1995). An important feature to highlight in the paleorainfall graph published by Bar-Matthews and Ayalon (2004) (see figure 5.3) is the generally low rainfall amounts throughout the entire Late Holocene. In spite of the fluctuations, the rainfall is significantly below that of today, with the exception of the periods from circa 200 BC to 200 AD, when it hovered around today's mean, and from circa 1400 to 1600 AD, when it reached a higher peak than the current average.

A complimentary $\delta^{18}O$ isotope record comes from cores taken in the Eastern Mediterranean Sea, just off the coast of Israel (see figure 5.5). Schilman et al. (2001, 2002) established that under present-day conditions, fluctuation in rainfall quantities is the major factor controlling changes in $\delta^{18}O$ in both the sea surface and the cave drip water. Allowing for slippage in the dates due to the complications of matching radiocarbon dates from the sea cores with the ^{230}Th-U dates from the speleothems, Schilman et al. (2002) conclude there were humid events peaking at circa 1200 BC, 700 AD, and 1300 AD, and arid peaks at approximately 100 BC, 1100 AD, and 1700 AD. The warm humid phase at 1300 AD corresponds with the globally important Medieval Warm period known from historical records and proxy climatic data throughout Europe and Asia (Lamb 1995; Roberts 1998), and the cool arid phase at 1700 AD can be correlated with the well-known Little Ice Age from around 1350 to 1850 AD (Lamb 1995; Fagan 2000).

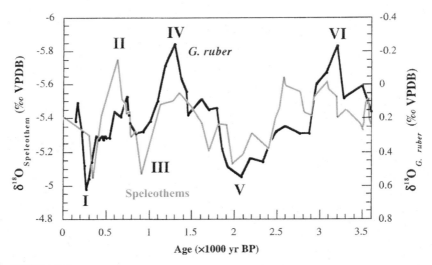

FIGURE 5.5
Comparison between the Marine Oxygen Isotopic Record of *G. Rubber* and the Superimposed Oxygen Isotopic Record of the Soreq Cave Speleothems for the Last 3600 Years (From Schilman et al. 2002).

It is clear that the highs and lows of humid and arid events in the Near East are important for understanding adaptations on the part of the city-states and empires in this region throughout the Late Holocene. However, a more important aspect of understanding the relationship between complex societies and climatic change is the *frequency* and *degree* of rainfall variability. This is alluded to by Bar-Matthews, Ayalon, and Kaufman (1998). Figure 5.3 is a revised version of this. It is immediately obvious from this phasing that the Middle Holocene was characterized by major fluctuations in humidity and very high amplitude change. This can be contrasted with the first part of the Late Holocene, which is the most climatically stable portion of the entire Holocene. The implications of this should be immediately apparent to students of climate and society. The degree of climatic fluctuation matters less than the frequency and predictability of the changes. Human societies are highly adaptable and can adjust their responses to climatic change if the change is relatively predictable. It is the unpredictable change that causes major stress to the fabric of a given society. Therefore, although the period from circa 1900 BC to 1000 AD had many shifts in wet and dry phases, these were at a relatively low magnitude, and therefore, much easier to adapt to than the shifts at the end of the third millennium BC Early Bronze Age.

Pollen Evidence

The most relevant pollen diagrams for the Late Holocene in the Levant are from Lakes Hula (Baruch and Bottema 1999; van Zeist and Bottema 1982), Kinneret (Baruch 1990) in northern Israel, Birkat Ram in the Golan (Schwab et al. 2004), and Ghab in northwest Syria (Yasuda, Kitagawa, and Nakagawa 2000). In Baruch and Bottema's (1999) Hula diagram the Late Holocene is fixed by the following dates, which are corrected after Cappers, Bottema, and Woldring (1998) and calibrated: zone 7, circa 950 BC (4140 ± 50 bp); zone 8, circa 375 AD (3280 ± 70 bp), and upper zone 9, circa 545 AD (3080 ± 70 bp) (see figure 4.4). The Late Holocene began at circa 2000 BC, which can be interpolated from the calibrated dates to a point near the top of zone 6. Shortly following this point there is a drop in overall arboreal pollen, probably reflecting an increase in human land use. A pronounced drop in arboreal pollen occurs at Birkat Ram after circa 400 BC. In zone 7 at Hula, an expanded olive peak could be associated with widespread olive production during the Iron Age. At this time the Southern Levant is littered with evidence for olive oil production including heavy stone presses, macrobotanical remains, and settlement evidence (Holladay 1998). In the middle of zone 8 there is a sharp decline in olive pollen, which corresponds to the cessation of olive horticulture sometime between the end of the Iron Age and the beginning of the Byzantine period. There is a slight rise in oak pollen at this time in response to the decrease in olive, indicating a recovery of the natural forest from the impact of human land use at that time. In the lower part of zone 9, the olive pollen begins to make a strong recovery marking the renewed extensive exploitation of prime horticultural land by the Byzantine period inhabitants of the region. This Byzantine period olive peak is mirrored in the Birkat Ram diagram as well (Schwab et al. 2004)

The Late Holocene portion of the Kinneret diagram begins in zone X_2, which is dated by Baruch to 1600–350 BC after correction and calibration. This zone indicates a decrease in arboreal pollen, which Baruch (1990) attributes to expanded settlement around the lake at that time. After this, in zone Y (350 BC–550 AD) there is a large rise olive pollen, which peaks within the Byzantine period as at Hula and Birkat Ram. The decline of the olive at the end of this zone is accompanied by an increase in oak and pistachio, again a sign of forest regeneration, possibly during the Medieval period.

In the Northern Levant, the Ghab pollen diagram from Yasuda, Kitagawa, and Nakagawa (2000) has only one date of circa 600 BC (after correction and calibration according to Cappers, Bottema, and Woldring 2002) for the Late Holocene, which falls in the middle of zone 5 (see figure 4.3 in the previous chapter). The general picture again indicates a major expansion of olive until zone 6 at the upper level of the core, which records a decline in olive and an increase in pine, suggesting reforestation at the higher elevations of the slopes.

Lacustrine Evidence

The Dead Sea is the source of a high-resolution lake level record for the Late Holocene, which has important implications for rainfall amounts for the last 4000 years. Comparisons of premodern records of lake level fluctuations and their relationship to Levantine rainfall patterns led Enzel et al. (2003) to describe the lake as a large rain gauge for the region. They determined that in the 115 years of rainfall records from the Jerusalem station, beginning in 1847 until the early 1960s before large amounts of water were drained artificially from the Dead Sea, high lake levels were associated with average rainfall amounts of 650 mm/year, and low levels corresponded to an average rainfall of only 445 mm/year.

Bookman et al. (2004) report on a systematic study of shore sediment deposits that preserve a record of lake-level fluctuations going back to around 2100 BC and compare these depositional environments to modern analogues within the range of Dead Sea deposits. They also acquired a large number of radiocarbon dates from a substantial amount of detrital wood embedded within the sediment units, thus allowing them to construct a detailed set of curves indicating level changes. Their evidence begins with sediments indicating a low lake level (around 411 mbsl) dated to circa 2140–1445 cal. BC (3703 ± 37 and 3220 ± 36 bp) (see figure 5.6). This represents a lake level that was receding from a previous high stand during the Middle Holocene. It corresponds with other evidence, such as that from the Soreq isotope record, for

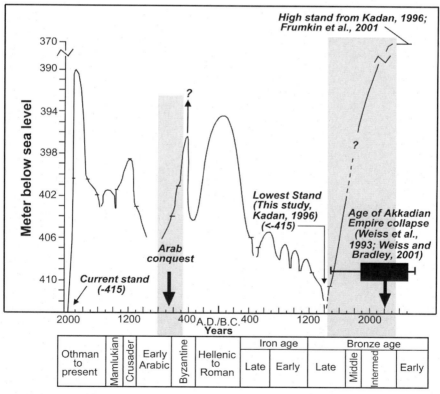

FIGURE 5.6

Summary of Evidence for Dead Sea Levels for the Late Holocene (After Enzel et al. 2003).

sharply declining rainfall amounts. Frumkin et al. (1991) also report evidence for increasing aridity at this time from the narrow size of the salt cave openings in the Sedom diapir.

Between circa 1000 and 550 cal. BC (2775 ± 37 to 2518 ± 35 bp), the lake levels fluctuated, indicating mildly wetter and drier episodes within an overall dry environment. There was a subsequent drop in lake level to 407 mbsl between 690 and 410 cal. BC (2518 ± 35 and 2485 ± 40 bp) with a very significant recovery up to 395 mbsl at a time period from circa 370 BC (2230 ± 30 to 1760 ± 40 bp) to 350 AD (2128 ± 27 to 1771 ± 30 bp). This high lake stand corresponded with the Hellenistic and Roman periods in the Levant, and the data are supported by Frumkin et al.'s (1991) evidence for wide salt cave openings in the Sedom diapir at that time, as well as archaeological evidence for second- and first-century BC boat moorings at a high elevation on the north shore of the Dead Sea (Bar-Adon 1989:3–4). After this Roman period high level, there was a rapid drop to 404 mbsl, and then the lake rose again to a short-term high stand dated to circa 400 AD (1630 ± 40 bp). After the fifth century AD, the lake level dropped rapidly to a low of approximately 413 mbsl. It recovered again only in the eleventh century AD (Bookman et al. 2004).

A portion of this scenario is also supported by evidence for a Dead Sea core published by Heim et al. (1997), who date deep water lake sediments to the Hellenistic through at least the Byzantine period. Klinger et al. (2003) report on a high stand of lake sediments deposited at the toe of the Wadi Hever alluvial fan dating to a time period between 50 BC and 240 AD.

Lake Kinneret in northern Israel provides a similar record of wet and dry episodes. Dubowski, Erez, and Stiller (2003) examined several short, subrecent cores from the lake and compared changes in carbonate chemistry to the environmental record since 1874. They then examined the carbon and oxygen isotopes from a core (KINU8) that had been collected from the lake in 1979 (Thompson et al. 1985). The dates on this core were corrected for the hard water error and calibrated. The results indicate that in the time period from circa 1300 to 600 BC (Late Bronze, Iron Age), there was a low nutrient flux and high $\delta^{18}O$ and $\delta^{13}C$ values, indicating more arid conditions. The period from circa 600 BC to 350 AD (the Persian through Roman periods) was marked by a return to a significantly moister environment as indicated by a sharp decline in the $\delta^{13}C$ and $\delta^{18}O$ values and a corresponding decrease in carbonate content. After this there is a hiatus in the sediment record, but it picks up again with a return to dry conditions in the period from circa 700 to 1050 AD (Umayyad and Abbasid Early Arab period). At 1050 AD there is a brief wet period and then a return to dry conditions until circa 1780 AD.

Geomorphological Evidence

Subsequent to the Middle Holocene alluvial episode, the period from circa 2200 BC to 0 BC/AD is primarily marked by phases of stability interspersed with episodes of channel downcutting. This is most likely related to the lowering of water tables and stream base levels as a hydrological response to the drier climate of the Late Holocene as indicated by the isotopic signatures of the period. There are a few exceptions with localized incidences of colluvial deposition associated with abandonments of agricultural terraces in specific stream valleys. One example of this is the Iron Age deposits recorded at Tel Lachish in the Shephela of central Israel. These date to the post-Assyrian conquest of the site, which led to abandonment of the town by the inhabitants and the resulting neglect of agricultural fields in the site vicinity (A. Rosen 1986b).

The alluvial hydrology changed relatively abruptly during the Roman and Byzantine periods in the Near East. At this time there was a shift from a regime of downcutting to one of alluviation. This alluvial episode was quite different in lithological signature from that of the Middle Holocene. Rather than an alluvial deposit with laminated facies indicative of well-sustained perennial flows, this fill is represented by massive fluvial silts in the northern Negev (Goodfriend 1987; Goldberg and Bar-Yosef 1990) and poorly sorted silt, sand, and gravel further north in the Shephela and Jezreel Valley (A. Rosen 1986c, 2006). This "Byzantine period" fill appears to have been deposited as a result of flash-flooding (A. Rosen 1986c, 1986b; Goldberg 1994). Similar alluvial fills from this time period appear in many localities from the semiarid zones of North Africa throughout the Near East as far north as Anatolia. They were first described as the "historical fill" by Vita-Finzi (1969), and many scholars since have associated them with evidence for erosive land-use practices during the Classical and Postclassical periods (Butzer 1980; Wagstaff 1981). These fills date to different time periods within the last 2000 years, and, therefore, they cannot be considered good climatic proxies. In the Near Eastern Mediterranean and desert zones, this latest phase of the Late Holocene is characterized by the most intensive land use of any time previously. Deforestation was widespread, along with agricultural terrace building and the practice of cash cropping vast tracks of land. These events contributed to widespread increases in the amount of soil erosion and hillslope wash leading to streams with higher sediment yields and several meters of valley fill deposited on older sediments.

In Israel, Jordan, and northern Sinai, this later historic-period fill tends to cluster into two episodes falling within the Byzantine and Medieval periods. The fill is widespread throughout northern and central Israel, identified as the top 2–4 m of alluvial valley sediment, and dated by ceramics to the Byzantine period or later (A. Rosen 1986c, 2006). Goldberg dated two occurrences of this fill to circa 300 AD (1755 ± 105 bp) and circa 1300 AD near Qadesh Barnea, Sinai (Goldberg and Bar-Yosef 1982). Cordova describes a lower terrace of gravel, silt, and sand lenses on the Madaba plateau of central Jordan containing Roman period (63 BC–324 AD) ceramics in the Wadi ath-Thamad, as well as a low 3 m terrace in Wadi Al-Wala with a sequence of silt beds separated by organic units dating to circa 850 AD (1190 ± 50 bp). The latter terrace suggests a stable wetland environment at this locale (Cordova 2000). Further north in the lower Balikh Valley of northern Syria, Wilkinson reported on 60–80 cm of alluvial silt/clay deposits with embedded sherds dating from late Roman–early Islamic periods. He also described aggrading floodplain environment during or after the late Roman period in the lower Orontes floodplain of southeastern Turkey (Wilkinson 1999).

It is clear that these Late Holocene alluvial fills are diverse in lithology and in timing. Due to this diversity they are poor proxy indicators of environmental change. Geomorphologists are hard pressed to prove that they are caused exclusively by shifting rainfall patterns on the one hand, or by erosive human land-use practices on the other. At the risk of taking a firm stand for indecision, it seems best to take the middle ground of this debate. On the side of climatic change, cave speleothems show that rainfall amounts during the period from circa 1000 BC to 1000 AD were relatively stable compared to times previously or since that period. However, within the range, there were episodes of greater and less rainfall, which would have had a significant impact on the large populations that lived in this semiarid zone. An upward shift in rainfall amount might encourage farmers to enter marginal farming zones, especially with the security of imperial Byzantine backing as a safety net. This would have led to greater amounts of plowing and more soil loosening on slopes. In the presence of increased rainfall, there would be larger amounts of sediment transported and finally deposited on valley floors, contributing to the alluviation in the stream systems.

SUMMARY OF HOLOCENE CLIMATIC CHANGE AND ENVIRONMENTAL SETTINGS IN THE NEAR EASTERN LEVANT

The above discussion of evidence for Holocene climatic change includes a complex mosaic of data from a wide variety of proxy evidence. Some of this evidence is complementary, and some seems to be contradictory. However, as more and more high-resolution data sets are added, we are becoming better able to identify some of the major trends in the story of environmental change and its impact on human societies in the Near East. This evidence is summarized in figure 5.7 and below.

Early Holocene Paleoenvironmental Setting (9500–5500 BC)

Isotope, pollen, lake-level, and geomorphological evidence from the Levant has given us a picture of the overall rainfall regime in the Early Holocene. The period began at circa 9500 BC with a steady amelioration toward a warmer wetter environment after the harsh episode of cool dry climate during the Younger Dryas. There was a continuous increase in rainfall, which was more evenly distributed over the year, including gentle summer rains. In the Mediterranean zone this led to a hydrology of active springs, high water tables, and streams with a steady, predictable perennial flow. Soils on the alluvial bottom lands were rich in organic material and sometimes waterlogged. This is exactly the type of environmental setting that would have provided a relatively secure crèche for the first established farming societies in the Near East (Sherratt 1980). Increased rainfall in semiarid areas often means more predictability in annual average rainfall amounts, and fewer drought years. This translates to lower risks for farming societies, especially if the high water tables were exploited for simple spring irrigation.

The hillslopes were well vegetated, but in the localities where Neolithic communities settled, the localized use of hillslope vegetation contributed to instability and slope wash. In the semiarid steppe and desert zones, the increased rainfall encouraged an increase in shrub vegetation and the spread of grasslands into desert regions. More burrowing animal activity led to an increase in slope wash and colluvial deposition. After about 7500 BC, the amplitude of rainfall fluctuations increased dramatically with relatively rapid shifts in $\delta^{18}O$ curves. In addition to regular small fluctuations, the early Holocene moist phase was punctuated by two of the wettest periods of the Holocene around circa 6400 BC and 5600 BC and an abrupt and short-lived

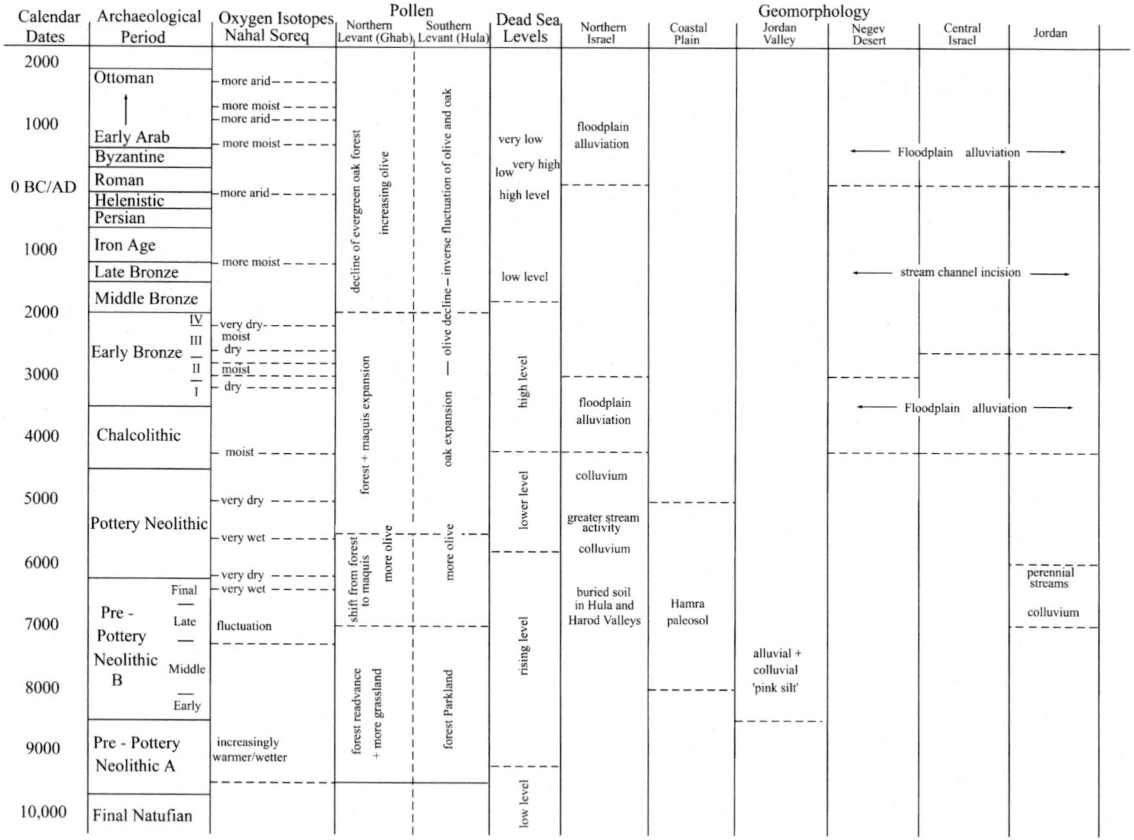

Chart content (rotated figure):

Calendar Dates	Archaeological Period	Oxygen Isotopes Nahal Soreq	Pollen — Northern Levant (Ghab)	Pollen — Southern Levant (Hula)	Dead Sea Levels	Geomorphology — Northern Israel	Coastal Plain	Jordan Valley	Negev Desert	Central Israel	Jordan
2000	Ottoman	more arid / more moist / more arid	decline of evergreen oak forest	olive decline – inverse fluctuation of olive and oak	very low / very high					Floodplain alluviation	
1000	Early Arab / Byzantine	more moist	increasing olive		low / very high / high level	floodplain alluviation					
0 BC/AD	Roman / Helenistic / Persian	more arid		oak expansion						stream channel incision	
1000	Iron Age	more moist			low level						
2000	Late Bronze / Middle Bronze	very dry-moist / dry	forest + maquis expansion		high level	floodplain alluviation				Floodplain alluviation	
3000	Early Bronze IV / III / II / I	moist / dry			lower level	colluvium					
4000	Chalcolithic	moist	shift from forest to maquis		lower level	greater stream activity					
5000	Pottery Neolithic	very dry	more olive	more olive	lower level	colluvium					perennial streams
6000		very wet			rising level	buried soil in Hula and Harod Valleys	Hamra paleosol				colluvium
7000	Pre-Pottery Neolithic B (Final / Late / Middle)	very dry / very wet / fluctuation	forest readvance + more grassland	forest Parkland							
8000	Pre-Pottery Neolithic B (Early)							alluvial + colluvial 'pink silt'			
9000	Pre-Pottery Neolithic A	increasingly warmer/wetter			low level						
10,000	Final Natufian										

FIGURE 5.7

Chart Summarizing Proxy Evidence for Climatic Change throughout the Holocene (Prepared by A. Rosen).

cool dry episode known as the 8.2KY event, which took place over a few hundred years from circa 6200 BC.

If we can rely on the dates for the Bar-Matthews et al. (1999) $\delta^{18}O$ and $\delta^{13}C$ sequences, the data indicate that the Pre-Pottery Neolithic A period took place in an episode that was becoming steadily warmer and wetter after the end of the Younger Dryas cold dry episode. The climate steadily improved throughout the subsequent early Pre-Pottery Neolithic B period with a continued trend toward moister climatic conditions. Sometime in the middle to late PPNB and PPNC, rainfall began to fluctuate between moist and dry conditions. These pronounced fluctuations continued into the Pottery Neolithic period. Bar-Yosef (2002b) has suggested that the very dry conditions associated with the 8.2KY event would have impacted populations living in the Levant in the PPNC and might have been largely responsible for the transitional nature of the period.

In the Northern Levant, the rapid shift to warm moist conditions allowed the evergreen oak forest to migrate rapidly into the Mediterranean zone, along with terebinth trees (*Pistacia* sp.). Conversely, the forests of the Southern Levant had a much less impressive recovery. It seems unlikely that this would be a function of strong climatic differences between the Northern and Southern Levant. Rather, this might be a result of significant forest and vegetation management on the part of early Neolithic populations, as suggested by Roberts (2002) for Neolithic Anatolia. However, the archaeological evidence suggests a relatively low village population at this time in the Southern Levant, and one would not expect there to be a need for extensive farming and widespread deforestation. If the slow forest recovery in the Southern Levant were indeed a case of early land management, it might instead represent an attempt to maintain a large open landscape to attract game closer to sedentary settlements, rather than a result of a broad farming base in the area.

Middle Holocene Paleoenvironmental Setting (5500–2000 BC)

The Middle Holocene opened at circa 5500 BC in one of the wettest phases of the entire 11,500-year period. This roughly coincided with the first half of the late Pottery Neolithic B "Wadi Rabbah" period (Kuijt and Goring-Morris 2002). The amplitude of climatic change was less than that of the Early Holocene, although significantly greater than that of the Late Holocene. Rainfall steadily decreased throughout the PNB, only returning to somewhat moister conditions within the Chalcolithic period. At that time there were moister episodes in the middle and then late Chalcolithic through Early Bronze Age I periods. More rapid-fire fluctuations in rainfall characterized the EB II and EB III, with generally wetter conditions near the end of the period. The Middle Holocene famously closed at the end of the third millennium BC with severe droughts that ushered in the generally drier Late Holocene.

Pollen evidence from the Mediterranean zone of the Northern and Southern Levant shows an expansion of oak and pistachio forest from circa 4500 BC. The forested hillsides existed throughout the Middle Holocene, suggesting that the impact of Chalcolithic and Early Bronze Age people on the landscape was probably restricted to localized areas around larger population centers. At the end of this period around 1900 BC, the oak forest declined in some areas of the Northern and Southern Levant. This could be another result of the drier climatic conditions that set in at the beginning of the Late Holocene, but it might also be an indication of human-induced deforestation.

The landscape changes during the Middle Holocene are an important part of the story of land use and economy for the Chalcolithic and Early Bronze Age societies that lived in this region. This is covered in detail in chapter 7. In the early stages of the Middle Holocene, much of the Mediterranean zone was carpeted in remnant red terra rossa soils, which were held in place by the relatively undisturbed vegetation cover. The valley sediments were black marshy deposits that retained moisture from the high water tables and perennial stream flow. At around 4500 BC the hydrological regime changed from a stable system to one characterized by increased stream energy resulting in valley alluviation and floodplain buildup. In the valleys of northern Israel, the sediments maintained their fine-grained waterlogged character. In central and southern Israel the streams deposited laminated silts and clay.

In both the north and the south, this change in stream regime provided a distinct advantage to the newly developing complex societies that lived in the region. The broadening flooding zones with renewed depositions of fine-grained deposits were ideal locations for floodwater farming. The opportunity to expand the zone of floodwater farming in the Chalcolithic and Early Bronze Age added a beneficial buffer to the farming economies, reducing risk and allowing the higher cereal yields needed to support growing populations in the ever-increasing numbers of new growing towns.

With the steady reduction in rainfall at the end of the Middle Holocene, the stream regime shifted from valley alluviation to valley incision. The opportunity for simple floodwater farming was removed from the farming equation, causing great stress to the populations living in the region at the time.

Late Holocene Environments

One of the most important aspects of the oxygen isotope curve for rainfall fluctuation in the Late Holocene is its lower variability in comparison to the curves for the Early and Middle Holocene. This is an extremely important factor, considering the role of climate change and its effect on the economies of the complex state societies and empires that existed in the Near East during this time period. The $\delta^{18}O$ curves from sea cores in the Mediterranean suggest moist intervals at roughly 1200 BC (around the Late Bronze Age), circa 700 AD (the Early Arab Umayyad period), and at the end of the Medieval Warm period circa 1300 AD (Late Arab Fatimid period). Dry episodes were identified at around 100 BC (late Hellenistic period), circa 1100 AD (Crusader period), and circa 1700 AD, late in the Little Ice Age.

Although the larger population concentrations in the region might have been more vulnerable to relatively small climatic shifts due to their size and greater demand for food resources, they were also more capable of overcoming the range of climatic fluctuations by dint of their technological advances and access to world markets. Larger population concentrations give rise to a faster rate of technological innovation and improved methods of communicating solutions to problems of recurrent drought and food stress. Likewise, extensive trade networks and the infrastructure of empires facilitated the transfer of subsistence resources from areas with more abundant surpluses to regions suffering failed crops. Therefore, although we can pinpoint phases of wet and dry episodes in the Late Holocene record, their impact on Near Eastern societies at that time has to be considered in tandem with a number of other historical, political, economic, and social events.

The Late Holocene is the period in which the landscape and vegetation are profoundly transformed from a generally unmanaged state to one that reflects both the ideology and economies of the growing populations of inhabitants (Wilkinson 2003:143). This is especially true in the late Roman and Byzantine periods, when the ideal of Mediterranean environment and economy was imprinted across the landscape even beyond the Mediterranean ecological zones, well into the semiarid steppe and desert lands (see chapter 8 of this volume). This is reflected in the Hula, Kinneret, and Birkat Ram pollen evidence, which shows a large spike in the olive curve during the Byzantine period in the Southern Levant, and a downswing in both evergreen and deciduous oak.

There were also major changes in landforms and hydrology. The almost universal downcutting of streams at the beginning of the second millennium BC was replaced at different times and in different localities by localized alluvial valley fills. During Roman/Byzantine times a more widespread valley alluviation appears to be a combination of extensive and erosive land-use practices in tandem with fluctuating rainfall patterns. However, Wilkinson points out that by Late Holocene times, there is a very complex cultural landscape in place in the Near East that would impact sedimentation and erosional patterns in different ways. Cultivated fields, cleared forests, and quarries would increase sediment supply to the valleys, whereas terraced fields would hold back the sediment on the hillslopes while they were actively maintained. High rates of sedimentation in northern Israel coincided with increased settlement in both the lowlands and uplands (Wilkinson 2003:148).

The story of Holocene environmental change was played out on the landscape with the major factors varying between steadily increasing rainfall amounts and changes in the magnitude of rainfall fluctuations. Vegetation and landscape responded to these climatic fluctuations, which presented both opportunities and obstacles to human cultural development in the region. The cultural response to these changing environments was far from predetermined and certainly not unidirectional or homogeneous. Social groups responded in different ways depending upon their levels of technological development, their political and social organization, and their unique cosmologies. By the Late Holocene, human populations themselves were affecting the environment in significant and irreparable ways. Throughout the Holocene the tempo of human impact increased hand in glove with the enhancement in our abilities to adapt to climatic changes. The following chapters present case studies of the role of climate change in some of the most momentous developments in the history of the Southern Levant. These developments came about as foragers, early complex societies, and empires came face to face with some of the problems and opportunities offered by Holocene climatic fluctuations.

6

From Hunter-Gatherers to Village Farmers: The Role of Climate Change in the Origins of Agriculture

BACKGROUND—THE BASIC CONCEPTS

Of all the causal explanations for the origins of agriculture, two main factors stand out as common themes proposed by researchers as prime movers. These concepts are *population pressure* and *climate change*. The two are sometimes cited as interrelated causes, but often climate change stands alone as a major force pushing ancient groups of hunter-gatherers into adapting new methods of plant exploitation that eventually lead to increased sedentism and a village farming way of life.

Chapter 1 of this volume opens with mention of one of the first researchers to propose climate change as a sole prime mover for the origins of agriculture, namely Raphael Pumpelly. Pumpelly's excavation of Eneolithic Anau in Turkmenistan was one of the first to reveal an early oasis farming community located in a region that can support little or no agriculture under today's dry climatic regime (Pumpelly 1908). This theme reflects Pumpelly's long-term collaboration with Ellsworth Huntington, who developed the perspective in part from his work in Turkmenistan with Pumpelly. Huntington's textbook *Civilization and Climate* (1924) became a landmark study for the school of *climatic determinism*. The notion that adverse climatic change was a key factor in the origins of agriculture formed the basis of V. Gordon Childe's (1926) "oasis hypothesis." Here he proposed that drying climatic conditions forced humans, animals, and plants to converge into close proximity around oases. This close contact allowed hunter-gatherers to become familiar with the ecology of potentially domesticable species, which in the course of time led to cultivation and domestication as a natural outcome.

At the time Childe was writing there was little proxy evidence for climatic change in the Near East. Most of his evidence came from work on the retreat of continental glaciers in Europe and speculations about the effects of Late Pleistocene and post-Pleistocene changes in the Near East (see H. Wright 1993). Early investigations of this problem were undertaken by H. E. Wright in the Zagros Mountains, in connection with the Jarmo archaeological project directed by Robert Braidwood (1951). At that time Wright (1960) identified evidence for low snow lines, suggesting parallels to the European glacial sequence for the Late Pleistocene. However, his evidence suggested that the major climatic changes of the terminal Pleistocene were too early to have had a major effect on the origins of agriculture. This work coincided with a paradigm shift in archaeology, and archaeologists began to concentrate on social factors that might have driven the change from foraging to an agricultural economy. At this time the concept of "climatic determinism" as an explanation for agricultural origins was looked upon with disfavor (Binford 1968; Braidwood 1961; M. Cohen 1977; Flannery 1969; Higgs and Jarman 1969).

From the 1970s there has been a concerted effort to elucidate our understanding of climatic change in the Near East (see chapters 4 and 5). From this work we now know that there were major climatic changes in the region at the time that hunter-gathering Epipaleolithic peoples in the Levant were exploiting wild cereals. We have also become aware of the possible impact of the Younger Dryas episode bringing significant drought and strong seasonality to the Near East during the Natufian period in the Levant (see chapter 4). This more recent awareness of Late Pleistocene/Early Holocene climatic changes has encouraged a revival of interest in exploring the role climatic change played in Near Eastern agricultural origins.

However, the course of archaeological inquiry is neither smooth nor unidirectional. The renewed interest in climate as a catalyst for agricultural origins has engendered disagreements about the impact of climatic change, and in turn the nature of the social mechanisms leading to early cultivation. Recent speculations about the role of climate in the origins of agriculture can be divided into two basic interpretations. The first proposes that the Younger Dryas period was responsible for droughts and reduction in the basic resources (including cereals) upon which the Natufian populations depended, thereby forcing new solutions to periods of food stress. This ultimately led to the beginnings of systematic cereal cultivation (Bar-Yosef and Kislev 1989; Goring-Morris and Belfer-Cohen 1998; Henry 1989; Hillman 1996). The second point of view comes from the perspective of *opportunism* (Bottema 2002; McCorriston and Hole 1991; Roberts 1991). It suggests that agriculture began under conditions favorable for the growth and spread of cereals and that local populations were optimizing their resource options.

Hillman (1996), Moore (Moore and Hillman 1992), and others (Hillman et al. 2001) have presented the most sophisticated reconstruction of vegetation patterns for the Northern Levant at the time of the Natufian occupation, which covers the critical time periods of the latest Pleistocene and Younger Dryas events. Their conclusions are based on the pollen evidence from key cores in the Fertile Crescent, as well as macrobotanical data from two excavation seasons at Abu Hureyra, a site with Epipaleolithic and Neolithic settlements close to the middle Euphrates in Syria. The initial settlement of Abu Hureyra took place shortly before and during the Younger Dryas climatic phase, during a period of abundance as trees were recolonizing the moister localities of the region (Hillman et al. 2001). The macrobotanical remains from this phase provide abundant evidence for a large variety of edible seeds and fruits. However, after about 300–400 years of initial occupation, arid conditions began to take effect with a resulting change in the plant exploitation patterns at the site. Edible seeds of drought-sensitive plants dropped out of use as well as forest products such as acorns. Later, large-seeded legumes such as lentils disappeared from the macrobotanical record. In marked contrast to the decline in these wild seed grasses, both wild wheat and rye increased at the site before they too began to decline. Although their numbers are small, Hillman et al. (2001) found grains of morphologically domestic rye from this time period and suggest that the disappearance of the natural habitat of cereals led people to begin cultivating rye and possibly wheat during the height of the Younger Dryas dry phase.

Other authors (Bar-Yosef and Belfer-Cohen 1992, 2002; Goring-Morris and Belfer-Cohen 1998) have also pointed to the Younger Dryas climatic reversal as a key factor stimulating the origins of agriculture in the Near East. They propose that the reduction in wild cereal stands associated with this dry episode eventually led Natufian populations to begin cultivating wheat and barley. They stress that the switch to cultivation involved a combination of both environmental and social factors unique to Natufian populations of the Fertile Crescent. These social factors are outlined as sedentism, social organization, and mode of subsistence. Bar-Yosef and Belfer-Cohen also point out that these factors could have varied with different Natufian groups as each one selected different options for adaptations to specific environments within the Levant. As climatic conditions deteriorated in the Late Natufian, some groups increased mobility while others maintained a sedentary existence. Some groups like the Harifians, a culture complex derived from the steppe zone Natufians and located in what is today southern Israel, adopted a strategy of increased mobility to take advantage of sparser resources over a broader region. Further north the sedentary groups solved the problem of diminishing resources by planting cereals. The latter solution had a profound impact on Late Natufian and early Neolithic societies, leading to greater social cohesion, territoriality, and eventually the introduction of public structures and changes in ritual activity (Bar-Yosef and Belfer-Cohen 1992; Bar-Yosef 1996).

Henry (1989, 1991) proposed a composite model of foraging behavior of Levantine hunter-gatherer populations long before the intentional cultivation of cereals. He maintains that the beginnings of complex foraging strategies among the Levantine hunter-gatherers was a function of increased Mediterranean woodlands and associated grassland habitats. With warmer climatic conditions beginning to take hold after the Late Glacial Maximum, the expansion of the woodlands to higher as well as lower elevations provided increased cereal and nut resources for the Epipaleolithic populations in the region. These abundant resources could be harvested over a three-to-five-month period by following the vertical elevation differences in ripening times. This allowed for abundant harvests for relatively long periods throughout the year. This type of labor-intensive "vertical economy" (Flannery 1969) encouraged population rise and a semisedentary settlement for the processing and storing of these food resources.

According to Henry (1989) two converging factors led Natufian societies to begin cultivating cereals. These were population pressure and a reduction in the resource base of nuts and wild grains. With the onset of environmental degradation in the Younger Dryas phase, the woodlands retracted along with much of the storable resource base of the Natufians. To survive, the Late Natufian populations of this region took to cultivation of cereals to supplement the declining resources. In the drier semiarid southern areas, Natufian populations increased their mobility to maximize the shrinking resource base, becoming what is recognized as the Harifian cultural entity. This was in marked contrast to the adaptive transformations taking place in the northern Mediterranean zones of the Levant. Natufian society of northern Israel proved to be less resilient than that of the south due to its increased emphasis on sedentism during the Early Natufian period.

McCorriston and Hole (1991), Herbert Wright (1993), and Bottema (2002) propose an "optimization" view; they make a strong case for a natural increase in cereals with the spread of grassland environments during the Younger Dryas period (H. Wright 1993; Bottema 2002) and the Early Holocene (McCorriston and Hole 1991). They suggest that resource availability *increased* and the exploitation of abundant stands of wild cereals led to cultivation of the plants. McCorriston and Hole draw on models of climatic change in the Mediterranean from the Cooperative Holocene Mapping Project (COHMAP) members (1988) to demonstrate the likelihood of intense seasonality at the end of the Pleistocene with an emphasis on very warm dry summers and very moist winters as the new Mediterranean environment began to replace the more continental Pleistocene climatic regimes in the eastern Mediterranean region. Ecologically, this phase of marked seasonality, along with human impact on the landscape, increased the ecotonal zones and favored the spread of annual grasses, which are better adapted to strongly seasonal climates.

Pronounced seasonality also provided an incentive to the further development of cultural trends already in place during the Epipaleolithic in the Levant. These include the use of grinding technologies (K. Wright 1994) and systems of storage as a solution to dry periods in which resources were scarce. The storage itself was one factor contributing to an increase in sedentism (Rowley-Conwy and Zvelebil 1989) and later to the need to increase production with growing populations. McCorriston and Hole also mention possible manipulation of the landscape in the use of firing at the forest edge. This process, although hypothetical, could have begun in the Epipaleolithic to attract deer to more open clearings. This would also have had an effect on the wild cereals, creating favorable conditions as annuals colonized these ecotonal areas. In chapters 4 and 5, I suggest that firing of forests might be one possible reason for the lack of forest recovery recorded at Hula in the Terminal Pleistocene/Early Holocene when other pollen diagrams from the region record a recolonization of forests after the retraction attributed to the Younger Dryas dry episode. Roberts (2002) has proposed a similar situation for Early Holocene Anatolia. McCorriston and Hole (1991) emphasize the interplay between environmental stress, new resource opportunities, the importance of preexisting grain-processing technologies as well as social institutions that were structured to adapt better to this stress.

Herbert Wright (1993) has supported McCorriston and Hole's position by outlining the climatic events that led to the beginning of a Mediterranean climatic regime in the Terminal Pleistocene and Early Holocene. He has stressed the importance of climatic factors operating from circa 17,000 to 10,000 cal. BP that were responsible for increasing seasonality at that time. During the Late Glacial Maximum the European ice sheets were responsible for the cold dry air that reduced precipitation in the Near East and created the cool dry winters of the Late Pleistocene. When the ice sheets retreated in the Terminal Pleistocene and Early Holocene, the strength of these cold fronts diminished, and winters became moister in the Mediterranean region. Simultaneously, the summer westerlies that had been active in the Late Pleistocene shifted northward, thus reducing precipitation in the summer. Herbert Wright (1993) maintains that with the glaciers gone from continental Europe, there was greater summer insolation increasing summer temperatures significantly in the Near East. During the Glacial Maximum summers were typified by cool, cloudy, and moist conditions, but in the Terminal Pleistocene and Early Holocene they became markedly warm and dry. All of these factors created a seasonality that was even more distinct than at the present time and gave rise to a true Mediterranean climatic regime. Wright interprets these climatic factors, and not the effects of the Younger Dryas event, as the cause of the decline in the forests evident in the Hula pollen diagram (Baruch and Bottema 1991, 1999). This would mean that rather than a cold dry Terminal Pleistocene episode, the Near East was impacted by strong seasonality with an emphasis on warm dry summers.

The difference in these two interpretations (environmental stress leading to new human adaptations versus opportunism with increasing abundance) is highly significant for the proposed climatic effects on vegetation in general, and most significantly on cereals. In the Younger Dryas stress model, both the forest and the grasslands would have retreated together, causing a serious resource crisis for Natufian populations in the region (Rossignol-Strick 1995, Bar-Yosef and Belfer-Cohen 1992). In the seasonality model, the forests retreated due to severely dry summers, but there was an increase in annuals, including cereals, which then spread into new habitats. Whatever the causes of this forest decline, whether due to the Younger Dryas or seasonality, the proposal of increased stands of cereals at this time seems to fit with the pollen data from Lake Hula (Baruch and Bottema 1999), which shows an increase in gramineae with the retreat of the forests. Bottema (2002) later emphasized this again and pointed out that models for Natufian cultivation based on decreasing grassland resources are flawed in that they conflict with results from the pollen analyses.

NATUFIAN AND EARLY NEOLITHIC ADAPTATIONS

The story of the first cultivators in the Near East has traditionally revolved around Natufian societies who inhabited the region from circa 14,500 to 11,600 cal. BP. These people were once thought to have been the first farming societies, an idea proposed by Dorothy Garrod (1932, 1957), who was the first archaeologist to excavate their occupation sites. Researchers now believe that the Natufians were complex hunter-gatherers who intensively exploited wild plants, including cereals, and small game animals (see Bar-Yosef and Valla 1991). However, since the Natufians directly preceded the first true farming communities of the PPN in the Southern Levant, researchers continue to search for the beginnings of wild cereal cultivation during Natufian times. To understand better the relationship between the Natufians and their environment it is necessary to outline some of the notable characteristics of Natufian culture and lifeways.

Archaeologists who are directly involved in excavations and analyses of Natufian material culture have pointed out that it is a mistake to look at the Natufian culture as a monolithic entity through time and space. Natufian society has distinctive characteristics that let us identify them as a group; however, there are marked spatial and temporal variations in settlement, subsistence, and material culture that allowed researchers to divide them into geographic zones as well as discuss Early, Late, and Final phases (Bar-Yosef and Belfer-Cohen 1989, 1991; Goring-Morris and Belfer-Cohen 1998; Valla 1998a). The main geographic zone in which Natufian society is best defined is the "core" or "central" zone in the Carmel and Galilee of present-day northern Israel (Stordeur 1981; Valla 1998a) (see figure 6.1).

The Early Natufians of this central area are notable for their trend toward increasing sedentism (Tchernov 1991). There is evidence that, as in the case of many hunter-gatherer societies, the Natufians engaged in a pattern of seasonal group dispersal and aggregation. Through time, the period of aggregation expanded until at least part of the group remained resident in larger settlements such as Ein Mallaha (Eynan) throughout the entire year. This was the first sedentary adaptation in the prehistory of the region and is very likely one of the key factors leading to eventual cultivation of wild cereals and perhaps other grasses as well (Bar-Yosef and Belfer-Cohen 1992; Valla 1998b).

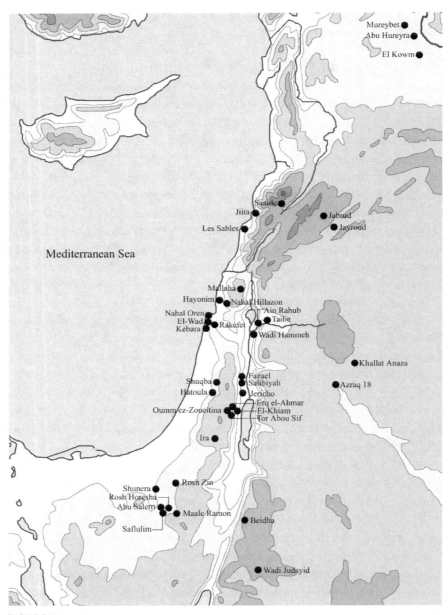

FIGURE 6.1
Map of Natufian Site Distribution in the Levant (Prepared by D. Beatty).

The origins of the Early Natufians are still not understood, although the new cultural traits associated with them began to appear around 14,500 cal. BP, after the post-LGM temperature rise (Bar-Yosef and Belfer-Cohen 1992). The Early Natufian (ca. 14,500–13,000 cal. BP) is characterized by a major increase in ground stone implements such as mortars, pestles, and grinding stones (K. Wright 1994). Architectural remains became more sophisticated with more carefully planned arrangements of semisubterranean structures than in previous periods (see figure 6.2) (Valla 1998a). Of critical importance is the appearance of cemeteries, which implies significant changes in social structure, territoriality, and belief systems. A major increase in art objects suggests new symbolic systems, and a much increased number of sickle segments in the lithic assemblages imply greater exploitation of cereals and other grasses. Sites such as Mallaha in northern Israel and Wadi Hammeh 27 in Jordan have led researchers to suggest an increase in sedentism as well as the social changes that would accompany such a shift in settlement patterns (Bar-Yosef and Belfer-Cohen 1992; Edwards et al. 1988; Perrot 1966; Tchernov 1991; Valla 1981). This assumption is supported by faunal evidence indicating long periods of residence at Mallaha. The faunal remains include commensals such as mice and rats (Tchernov 1991) and gazelle bones, indicating year-round hunting (Lieberman, Deacon, and Meadow 1990).

In the Late Natufian period (ca. 13,000–11,600 cal. BP) there appears to be a trend toward more mobility in the core Mediterranean zones, as indicated by smaller and less substantially constructed residences (Goring-Morris and Belfer-Cohen 1998). Subsistence practices still focused on collecting wild grains and nuts, as indicated by the continued presence of sickle blades and an increase in the number of sites with querns and mortars. There was also continued hunting of gazelle with the addition of larger numbers of waterfowl and other birds. At this time there appears to have been an expansion of Natufian influence with sites in the central core area moving down onto the lower elevations. An expansion to the northeast into modern-day Syria extended as far as the middle Euphrates. In the more arid zones such as the Negev and Judean Deserts, high altitude sites became more numerous (Goring-Morris 1998; Valla 1998a). In the Negev Desert, the settlement pattern became one of communities aggregating in the highlands during the spring/summer seasons, and dispersing into numerous small groups in the lowlands during the winter season (Goring-Morris and Belfer-Cohen 1998).

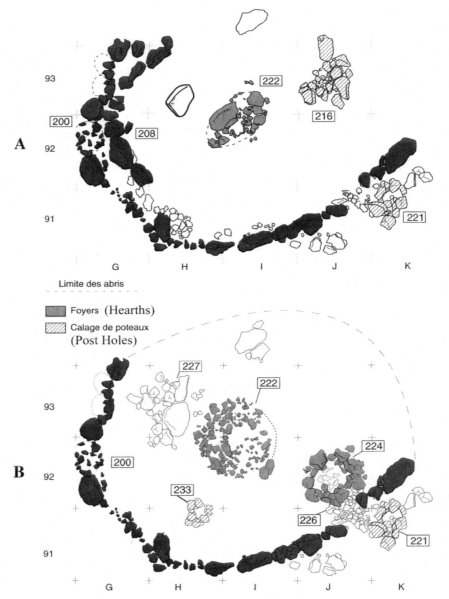

FIGURE 6.2
Plan of Natufian Structures from Ein Mallaha (After Valla et al. 2004, Figure 17).

The latter part of the Late Natufian period (sometimes distinguished as the Final Natufian period) occurred within the arid Younger Dryas episode. Significant changes appear to have taken place among Natufian populations of the Levant at this time. There was a major depopulation of the center core of the Galilee and Carmel areas, with sites such as Hayonim being abandoned. At Mallaha there was a decline in architecture. There was a clear disruption to the subsistence focus on wild cereals and other intensively collected plant foods. In the Northern Levant, the site of Mureybit continued to be occupied with a subsistence emphasis on wild cereals, although most other sites had been abandoned (Valla 1998a). Any tentative beginnings of cultivation that might have taken place at the height of Natufian occupation seem to have been stifled in the terminal phase of the Natufian cultural phenomenon.

The succeeding period of the Pre-Pottery Neolithic A in the Southern Levant (from ca. 11,700 to 10,500 cal. BP) first appears at sites in central Israel (Hatula) and the Jordan Valley. Its earliest phases overlap with the later Younger Dryas. The later PPNA spans the period from the end of the Younger Dryas through the onset of the Holocene. It is characterized by the first true villages such as Jericho, Netiv Hagdud, Dhra, and Wadi Faynan 16 in well-watered localities near springs and ponds, primarily in the Jordan Valley (Kuijt and Goring-Morris 2002).

The issue of the first sustained cultivation in the PPNA period is a subject of ongoing debate. Although Kislev (1999) and Daniel Zohary (1999) now agree that wild rather than domestic barley was exploited at Netiv Hagdud, there is still an ongoing pursuit to find the first appearance of the earliest morphologically domestic grains. Wilcox (2004) reports on increasing sizes of cereal grains from earlier through later PPNA levels at sites on the Syrian Euphrates. He believes this indicates the transformation from wild to domestic cereals. Colledge, Conolly, and Shennan (2004) summarize the evidence, and although it is controversial, they accept the possibility that there was early wheat and barley cultivation at the PPNA site of Iraq ed-Dubb in Jordan (ca. 11,500 cal. BP). Previous claims of early domestic cereals from Aswad (Level Ia) and Jericho (Hopf 1983) were erroneous based on recent re-dating of these levels to the subsequent PPNB period. Some researchers question the reliability of the evidence for PPNA cereal domestication based on the difficulty of distinguishing between wild and domestic cereal remains, and they also question the integrity of stratigraphic provenience or radiocarbon dates (Nesbitt 2002). However, there still remains the question of PPNA cultivation of morphologically wild grains. Arguments in favor of this are based on the composition of weed flora, as well as distance from wild stands (Colledge 1998; Hillman et al. 2001). However, new phytolith data from Netiv Hagdud, Dhra, and Wadi Faynan 16 suggest that although cereals appeared at these sites in the PPNA, they were not cultivated at these localities and were either cultivated or more likely collected at higher elevations (Rosen and Jenkins, in preparation).

The linking of climatic events with the transformation to domestication obviously relies on the precise timing of events. Unfortunately, the dating of climatic episodes is sometimes much less precise than that of the cultural transformations. However, if we look at the climatic trends as outlined by the Soreq Cave isotopic data, we see that the warming and wetting-up period immediately preceding the Younger Dryas episode seems to correspond with what we perceive as the florescence of Natufian culture including aggregation, architecture, artistic expression, elaborate burial rituals, and what may be the development of storage facilities. The full Younger Dryas climatic degradation generally corresponded to the decline of the "classic" Natufian way of life in the Mediterranean zone of the Southern Levant. The reversal of the drying trend toward renewed amelioration appears to have occurred at around 11,500 BP (U-Th. dates) according to the Bar-Matthews et al. (1999) isotope data from Soreq Cave. This coincides with a return to a more settled lifestyle at later-period PPNA sites. Although there is some evidence for the use of cereals at this time in the Southern Levant, there is very little secure evidence for true cultivation. At this time, the PPNA populations resumed the Early Natufian strategy of population concentration around water sources and intensive exploitation of stationary resources after the Final Natufian decline. Goring-Morris and Belfer-Cohen (1998) suggest that this renewal of an old adaptive strategy was a hedge against the rapid climatic fluctuations of the previous Final Natufian and early PPNA periods.

Settlement around permanent water sources and exploitation of storable resources such as cereal grains to buffer the community against lean years contributed to the success of this adaptation. Bar-Yosef further proposes that due to population constraints and increasing territoriality, the collection of these storable resources was geographically restricted, and, therefore, cultivation became a necessary strategy (Bar-Yosef 1998). Given the low recovery rate of the Early Holocene forest, as indicated by the Hula pollen data, it might also be possible that PPNA populations were burning forests to increase grasslands for denser stands of wild cereals and to encourage wild game, as suggested in chapter 5.

Although the above models are all informative about the factors that led the Natufian peoples to begin intensively exploiting wild cereals and other grasses, most of these depend upon an implication of continuity between Natufian adaptations and Neolithic village farmers. However, the link from Natufian subsistence to Neolithic villages is neither obvious nor direct in the archaeological record of the Southern Levant. Sites from the earliest Neolithic period in the Levant, the Pre-Pottery Neolithic A period, are sparse, and when they are found, there is little direct evidence for the cultivation of cereals. Unequivocal evidence for cultivation is rare even with the later PPNA, as discussed below, and true, unmistakable farming societies became significant only during the PPNB period (after 10,500 cal. BP).

In fact, the subsistence focus of the Natufians was in many ways similar to that of their late Upper Paleolithic/early Epipaleolithic period predecessors, who intensively exploited wild grain around Lake Kinneret some 8000 years before the Natufians came into being (Kislev, Nadel, and Carmi 1992), although the order of magnitude was much greater in the Natufian period. So rather than attempting to establish a direct causal link from complex hunter-gatherers to true village farmers in the Pre-Pottery Neolithic period, it is informative to try and understand two of the large steps along the way and the role climate played in these adaptations. The first step is the Natufians as foragers adjusting their lifeways and economies to the dramatic climatic changes of the Younger Dryas, and the second is the impact of Early Holocene climatic amelioration on the earliest village dwellers at the start of the Holocene, and its role in the shift from exploitation of abundant wild resources to an economy based largely on farming. To understand the role of climatic change in influencing the subsistence decisions of hunter-gatherers, it is important to consider some of the general principles governing the economies of these societies in marginal environments.

BUFFERING AGAINST FOOD
SHORTAGES IN HUNTER-GATHERER SOCIETIES

As astutely pointed out by Halstead and O'Shea in their 1989 introduction to *Bad Year Economics*, the only constant in human societies and their environments is the fact that there is continual change. Ethnographic studies of pre-modern societies often leave the reader with the impression that a given social system is a static entity situated within an equally static environmental milieu. Archaeology has given us the benefit of time depth. This diachronic perspective can function as a "time laboratory" within which to examine ideas and models generated by the ethnographic data. Conversely, in attempting to understand the motivations of prehistoric peoples it is essential to examine possible analogies from the ethnographic record. This is also true for investigations into the origins of agriculture in the Near East, where it is useful to frame the problem within the context of hunter-gatherer studies.

Much has been written on the issue of foraging strategies and the ways in which hunter-gatherers buffer themselves against environmental stress. Low (1990) proposed that hunter-gatherers respond to weather extremes rather than the average weather conditions. Since many environmental reconstructions are based on proxy evidence, such as pollen (see chapter 2), which respond over the long term to mean rainfall and temperature values, we must use caution when modeling the responses of hunter-gathers to environmental stress by attempting to differentiate between mean conditions and variability, as well as between long-term secular change and abrupt events. Low also maintains that foragers prefer to smooth over episodes of booms and busts by selecting resources with a lower mean payoff but greater stability. So rather than a feast and famine strategy, hunter-gatherers typically emphasize foods that are readily attainable and dependable, thus minimizing risks in times of environmental extremes. The Natufians, faced with the abrupt environmental change of the Younger Dryas, would be a good test case for forager responses to rapid environmental change.

Winterhalder and Goland (1997) take an evolutionary ecological approach to the role of climate in early domestication. In doing so they caution against taking "climate change" as a generalized element to which groups of foragers must adapt in an abstract sense. They suggest that environmental change should be broken down into its elements according to the resources impacted by this change. Such elements include the pattern of environmental variability, its frequency, duration, magnitude, spatial scale, and predictability. Given this, they suggest that researchers should examine the effects of changing resource parameters vis-à-vis a group of foragers in a well-defined environment with an explicit system of resource ranking. Winterhalder and Goland stress that prehistoric individuals did not face decisions in terms of climate shift, or mean annual changes; more important are the local factors such as the constraints a group faces, its actions, and the consequences of those actions (1997:126). The principle of "resource choice" is central to their model, which ranks resources according to abundance and preference. If a resource with a low ranking of preference, but high density, switches to a high ranking, then it will set mechanisms into motion toward intensive exploitation and eventual domestication, which ultimately become irreversible due to population growth and social investment. This follows through to models proposed by Flannery (1969), and later Rindos (1984).

To understand subsistence changes among foragers, it is necessary to recognize some of the risk-buffering mechanisms employed to overcome periods of environmental stress. These are outlined as five options: (a) resource selection, (b) intraband (and interband) sharing, (c) regional movement, (d) short-term carryover averaging (or short-term storage), and (e) long-term storage. They maintain that "sharing" and "regional movement" are most effective. Furthermore, Slobodkin and Rapoport (1974) suggest a hierarchy of responses to times of stress and environmental change including (a) speed of response, (b) degree of resource commitment, and (c) the reversibility of the response. The more slowly activated responses are less reversible and more "inclusive" of the greater social network.

All of the above are important considerations in understanding hunter-gatherer responses to economic stress. One might argue, however, that these concepts taken from the "optimal foraging" school of thought are too mechanical and do not take into account the belief systems and cosmology of a particular foraging group that may in fact direct a particular hunter-gatherer society down pathways of apparently nonoptimal solutions. Minnis (1985) points out that specific adaptations depend upon the unique social system, history, political situation, and technology of a particular group. This would account for some of the variability found in the response of different foraging groups to similar kinds of environmental stress within equivalent ecozones. Minnis emphasizes two common ways that preindustrial households diversify their resources to buffer themselves against periods of stress. As Winterhalder and Goland (1997) also say, these include (a) increased mobility to obtain access to a broader resource base, (b) utilization of a wide variety of secondary resources or "famine foods," and (c) the most effective solution, which is enlargement of the social networks. Minnis outlines various levels of response to resource stress that progress toward increasing social inclusivity. He stresses that "most human adaptive behavior is cultural" (Minnis 1985:20). Buffering can also come through technological change such as more efficient hunting and gathering techniques or the use of new tool types.

COMPLEX HUNTER-GATHERERS AND CLIMATE CHANGE—CASE STUDIES
Much important work has been conducted on Natufian sites in the Levant and
surrounding regions, and the relationship between climate change and the beginnings of cultivation have been pondered from several different perspectives
including both the *push factor* related to adverse climatic changes driving populations toward an agricultural economy, and the *pull factor* resulting from
environmental conditions favorable for wild cereals attracting populations to
more intensive exploitation and the eventual switch to agriculture. Before we
progress further in the discussion of the complexities of Natufian adaptations
in an era of significant climatic change, it is helpful to look at case studies
from locales with similar environmental constraints and parallels in sociocultural development among complex hunter-gatherer societies. These case studies are also based on archaeological data and therefore should be taken as
models subject to the same scrutiny as the Natufian models.

California Model
One such region is California. This area of the southwestern coastal United
States has an environment that shares many characteristics with that of the
Levant, most notably a true "Mediterranean" climatic regime defined by cool
moist winters, warm dry summers, and frequent occurrences of severe
drought situations. The vegetation ranges from pines in the high mountain
zones to oak-dominated Mediterranean chaparral and dry-land shrubs in the
desert regions.
Unlike the Natufians, traditionally some of the native Californians had a
subsistence focus on marine resources with a later emphasis on acorn collection after ca. 3000 BC. However, like the Natufians these groups could be clas-
sified as broad-based hunter-gatherers with complex social organization and
varying degrees of sedentism. McCorriston (1994) has pointed to these similarities, suggesting there is much one can learn from examining the subsistence trajectories through time of these two peoples.
Kennett and Kennett (2000) make a convincing case for the role of climatic
fluctuations in the increasing social complexity of the native Californians
from 500 to 1300 AD. They provide new, well-dated oxygen isotope data from
sea cores off the coast of Santa Barbara with a remarkably high resolution of
25-year intervals for the last 3000 years and 50-year intervals for the rest of the
Holocene. These data indicate that in contrast to Early and Middle Holocene
temperatures, the Late Holocene was characterized by much greater instabil-

ity in warm-cool temperature fluctuations. They found that in the period from 1050 BC to 450 AD, sea surface temperatures were relatively warm and stable. Then from 450 to 1300 AD a series of the coldest Holocene episodes ensued accompanied by unstable climatic intervals. After 1300 AD water temperatures returned to warmer and more stable levels. In comparing marine and terrestrial climatic records Kennett and Kennett show that cool marine conditions between 450 and 1300 AD correlate with a dry terrestrial period.

Even more significantly Kennett and Kennett show that these cold dry intervals from 450 to 1300 AD were characterized by high marine productivity and decreased terrestrial resources. The fluctuation in resource availability and the increase in unpredictability during this critical time period led to subsistence stresses that resulted in the development of new competitive as well as cooperative strategies. The marine resources were a focus of intensive exploitation as a more predictable food item than terrestrial sources. The diminishing supplies of perennial fresh water might have led to populations claiming and settling around springs and streams. The ultimate social consequences of these developments included sedentism, intensive fishing, greater social interaction, increased violence, and production of nonfood trade items. This situation did not lead to agricultural village life in the case of the pre-Columbian Californians, but, as McCorriston (1994) points out, their cultural trajectory was interrupted by the arrival of the Europeans in the sixteenth century AD.

The elements of this case study that most apply to the Natufian-through-PPNA transition are the cold dry conditions with increased unpredictability of terrestrial resources, the intensification of the more productive resources, and the eventually sedentary settlement around permanent water sources. In the case of the Levant, the Younger Dryas provided the environmental stress with its cold dry climatic conditions leading to a decrease in forest resources in favor of more productive steppic and grassland products.

Northern Australia
Moving significantly farther geographically, Haberle and David (2004) have investigated the role of abrupt climatic change in the shift from foraging to agriculture in northern Australia and the New Guinea highlands. Although this region is characterized by a very different ecology from that of the Near East, their research highlights some general principles of how hunter-gatherers respond to conditions of drought and stress that are useful for understanding

similar reactions on the part of Near Eastern Natufian and PPNA populations. In southeastern Cape York, they identified major regional demographic changes and intensification of subsistence practices that broadly corresponded to decreasing rainfall levels and increasingly open vegetation after 3500 cal. BP. Seed-grinding stones began to appear after 1800 BP, implying major changes in diet at that time. The increasing exploitation of the grassland areas transformed the relationship between people and their environment, presumably impacting how people perceived and symbolized their surroundings, as well as influencing task scheduling on a daily and seasonal basis. It also would have altered the way people used and managed their surroundings. These changes led to more concentrated use of marginal environments, increased landscape management including the use of intentional burning, and a broader subsistence base.

Social factors came into play such as fissioning of populations into smaller groups. These groups had a heightened sense of territoriality as shown by a regionalization of rock-art styles after 3700 BC. The result was more intensive use of plant resources within smaller areas, including seed grinding and a broadening of the types of plants exploited in these areas, resulting in such phenomena as the use of toxic plants not previously consumed. Finally, Haberle and David (2004) suggest that the transition to this late broad-spectrum adaptation was partly a result of climatic shifts, and partly a result of social agency. The latter factor is subject to a wide range of possibilities contingent on the particular society in question.

North American Great Plains

Responses to drought and stepped climatic degradation were also recorded for the North American Great Plains during the Middle Holocene (Meltzer 1999). Here drying climatic conditions led to a large reduction in populations of bison, the disappearance of surface and groundwater sources, a reduction in the density of vegetation cover, and a shift to more xeric grassland vegetation. These events promoted significant adaptive changes among the resident hunter-gatherer populations, although adaptations varied among different social groups. Some general trends included an expansion of the diet to include lower-ranked/higher-cost resources, the introduction of new technologies, relocation of settlements to better-watered areas, a trend toward more permanent settlement near available water sources, and a decrease in the number of sites, possibly representing reduced population levels. Archaeological evidence suggests that extended periods of drought on the Great Plains led to adaptive strategies that called for increasing technologies for intensive seed exploitation, processing, and storage, as indicated by larger numbers of grinding stones, earth ovens, and storage pits. In addition to this, settlement evidence shows that all the known archaeological sites for this period are located near perennial sources of water and indicate possible clustering of populations around these more resource-secure localities.

NATUFIANS REVISITED

Turning back to the case of human adaptations to climatic change from the onset of the Younger Dryas through the Early Holocene, we can now put Natufian solutions to problems of environmental stress into the framework of complex hunter-gatherers in similar situations worldwide, to compare and contrast responses between these societies. Tables 6.1 and 6.2 outline some of the key elements of hunter-gatherer adaptations to environmental stress and the responses of the social groups discussed above. We can now ask, what are some of the elements that led the Natufians to increased exploitation of wild cereals and their descendants to the adoption of agriculture, when in other

Table 6.1.   Hunter-Gatherer Buffering Devices and Responses to Climate Change.

Risk-Buffering Strategies
    Resource selection.
    Intraband sharing.
    Regional movement.
    Short-term carryover averaging (or short-term storage).
    Long-term storage.

Resource Choice
    Under environmental stress a resource with a low ranking of preference but high density switches to a high ranking; then it will set mechanisms in motion toward intensive exploitation and sometimes domestication.

Diversification of Resources
    Increased mobility to obtain access to a wider resource base.
    Utilization of a wide variety of secondary resources or "famine foods."
    The most effective solution is enlargement of the social networks, increasing social inclusivity.

Important Considerations in Understanding Responses to Times of Stress and Environmental Change
    Speed of response.
    Degree of resource commitment.
    Reversibility of the response. The more slowly activated responses are less reversible and more "inclusive" of the greater social network.

Table 6.2. Case Studies of Hunter-Gatherers under Stress from Environmental Change.

Case Study	Environmental Changes	Changes in Resource Choice	Settlement Changes	Intergroup Interactions
Californians	Warm moist to cool dry; increase in marine resources; decrease in terrestrial resources; less water available	Increased use of more predictable resources: use of more marine and less terrestrial; *diversification* acquired through trade rather than mobility	Increased sedentism to claim good fishing areas and fresh water sources	Increase in competition and hostility, but also increasing cooperation and trade
N. Australia	Decreasing rainfall; greater steppic vegetation	Broader subsistence base— more diverse resources used in smaller area; *labor-intensive* seed grinding and use of toxic plants	Breaking up into smaller groups; greater territoriality	Increase in territoriality
North American Plains	Drying conditions; fewer water sources; reduction in primary resource: bison; increase in grasslands	Increase in lower-ranked/higher-cost resources; new food-processing technologies such as grinding stones for grasses	Relocation to permanent water sources; more permanent settlement; fewer sites; more evidence for storage	
Early Natufians (ca. 14,500–13,000 cal. BP)	Warming and moist; increasing forests, including *Pistachia* and *Quercus*	Use of forest resources such as acorn and pistachio[6] (primary?); and grass seeds (secondary?); lower subsistence risks;[1] possible storage in baskets[6]	Increase in population in the Mediterranean zone; increasing sedentism and relatively large hamlets; architectural innovations; more sickles; more groundstone; more diverse toolkit; art	Restricted territorial ranges; well-developed exchange networks[2]

Period	Climate	Subsistence	Social	Trade
Late/Final Natufians (ca. 13,000–11,600 cal. BP)	Drying climate with Younger Dryas; decrease in forest, increase in grassland[7,8]	Possible decrease in forest resources such as pistachio and acorn; increase in grinding stones and sickles for harvesting in grasses; burning of land to maintain grasslands?[5] possible storage of seeds? increase in water fowl, depletion of gazelle	Decrease in sedentism and dispersal of smaller population groups; toolkits still include sickles and an increase in groundstone mortars[3]	Increase in territoriality; trade from the south as well as the Mediterranean[4]
Early PPNA (PPNA from 11,700 to 10,500 cal. BP)	Continued dry cool environment;	Continued use of wild grasses; greater exploitation of water fowl; use of fish	Continuation of mobile communities; sites around springs and permanent water sources; populations begin to grow	Trade in fish
Later PPNA	Warm wet Early Holocene; forests increase only slightly; perhaps grassland maintained by burning?	Intensive use of wild cereals; Planting of wild cereals and first truly domesticated cereals near the end of the period; continued investment in grass-processing technologies?	Full village life; rapidly growing population; first real evidence for silos and grain storage; rise in intensity of occupation[9]	

[1] Goring-Morris and Belfer-Cohen 1998
[2] Bar-Yosef 1996 [in humans at the end of the ice age]
[3] K. Wright 1994
[4] Bar-Yosef 1991
[5] See chapters 4 and 5, this volume
[6] Bar-Yosef and Belfer-Cohen 1992:32
[7] Baruch and Bottema 1991
[8] Bottema 2002
[9] Munro 2004

similar localities, there was no primary development of an agricultural lifestyle?

During the pre–Younger Dryas period, woodland expansion allowed the Early Natufians to take advantage of the forest resources in the core geographic zones. Although we have little direct evidence for plant food remains at Natufian sites, it is likely that some of the mortars at Early Natufian sites were used for processing acorns and pistachio nuts (Bar-Yosef and Belfer-Cohen 1992). This is supported by plant remains from the much earlier late Upper Paleolithic/early Epipaleolithic site of Ohalo (Kislev, Nadel, and Carmi 1992; E. Weiss et al. 2004). When given a choice, hunter-gatherers rank nuts at a higher level than grass seeds due to the ease of collection and high nutritional value (Gremillion 2004; Winterhalder and Goland 1997). Cereals and other large-seeded grasses can be an important part of the diet as well, although as a secondary resource. Since nut-producing trees have variable yields from year to year, grass seeds are known ethnographically to fill in for the years in which nut collection is not productive. They are therefore a critical, although not necessarily a primary resource.

The lush environmental setting of the Early Natufian period gave way to the cooler drier Younger Dryas with a decrease in tree cover and an increase in grassland in the Natufian core area (Bottema 2002). Again, studies of forager communities reveal that with the onset of environmental stress, one of the first strategies of hunter-gatherer groups is to switch from the higher-ranked resources requiring less relative effort to the lower-ranked commodities that are more abundant or more predictable but for one reason or another are less desirable. This switch is undertaken even though the lower-ranked items might require a significantly greater investment in time and energy for collection, capture, and/or processing. They are also likely to require the development of new technologies for exploitation and the implementation of new settlement patterns. With cooling and drying conditions a common strategy is to go after more predictable, high-yield resources, as in the case of the Californians. If this is impossible, then hunter-gatherers diversify and spend more effort on secondary resources, as in the cases of the Australians, North American Great Plains hunters, and the Late Natufians. All three of these groups took advantage of the expanding grassland habitats to exploit grasses at the expense of greater time, effort, and in the case of the Australians and North Americans the development of new grinding technologies. For the Late Natufians, grass seeds were already a part of the diet. Mobility decreased for the Californians and the Plains dwellers, who focused on stationary but renewable resources and permanent water sources. Mobility increased for the Late Natufians and Australians.

DISCUSSION

Late and Final Natufian society is often portrayed as a community undergoing resource stress, with this condition giving way to new adaptations, primarily that of wild cereal cultivation. This is partly based on the assumption that the Younger Dryas dry episode resulted in the reduction in the stands of wild cereals. However, evidence from Lake Hula points to an increase in grasslands at that time with a decrease in the woodland zones (Baruch and Bottema 1991; Bottema 2002). Therefore, the most important environmental stress of the period appears to be a reduction in the Mediterranean forest zone and a possible decrease in permanent water sources.

According to the model outlined in the preceding section, the readjustment of the Late and Final Natufian subsistence focus away from tree products, and toward a grassland exploitation strategy in the Younger Dryas, appears to have been a very successful adaptation. Furthermore, rather than simply a time of stress, it is more likely to have been a readaptation that allowed Natufian societies to continue to exist as complex hunter-gatherers in the Southern Levant for approximately 1500 years with a measure of stability. Munro (2004) also points this out, reaching a similar conclusion based on the small-game exploitation patterns in Natufian sites.

In the quest for understanding the origins of food production in the Near East perhaps it is a mistake to focus primarily on Natufian adaptations. For one thing, with the exception of Abu Hureyra on the Syrian Euphrates, we have little direct evidence that the Natufian peoples were planting wild cereals rather than collecting them. Even in the succeeding PPNA, there is more evidence for wild cereals rather than morphologically domestic ones (Harris 2003; Nesbitt 2002), and this leads to the possibility that even at that time the populations continued to rely heavily on collected rather than cultivated grains. Given this situation, it might be more reasonable to see the actual "threshold" for the origins of farming communities not in the Natufian period but rather in the later PPNA or even early PPNB, well into the Early Holocene warm wet phase. Rather than looking at *push factors* and *adaptive pressure* from degrading climate as the incentive for the origins of agriculture, it might be more productive to seek out a social *push factor* in the form of growing populations in the later PPNA and a climatic *pull factor* in the form of an ameliorated Holocene climate.

Agriculture is a very high-risk venture. It makes much more sense that the investment in an agricultural lifestyle would be undertaken in an environmental setting that was predictable and in which there was a lower risk factor. If we look at cases for agricultural dispersal, societies typically adopt agriculture under more favorable climatic conditions. This is evident from the first agricultural communities in central Asia during the Middle Holocene *climatic optimum* and also evident from the rapid spread of the agricultural way of life during the warm moist PPNB period in the Levant. Agriculture disperses into new areas under favorable conditions, not in times of stress.

CONCLUSION

There can be little doubt that the abrupt and rather dramatic climatic changes that occurred in the Younger Dryas at the terminus of the Pleistocene would have had a major impact on hunter-gatherers living in the Near East at the time. However, the question remains whether this impact was a prime mover toward the first steps to a settled agricultural life. As discussed above, when faced with periods of environmental stress, hunter-gatherers employ a variety of options including changing emphasis on different resources. The question then becomes (a) were the subsistence changes activated by Late and Final Natufian population directional and irreversible, leading to the eventual and inevitable origins of agricultural village economies, or (b) did the Natufians in fact reach a kind of stable equilibrium in their resource exploitation strategies? If we adopt the former perspective, then we must also ask why there was no switch to an agricultural lifestyle during earlier phases of climatic change, such as during the period of cold dry conditions associated with the Late Glacial Maximum. This is the time period when late Upper Paleolithic/early Epipaleolithic foragers at Ohalo II were exploiting steppic grasses in the same manner as the later Natufian populations.

This emphasizes the point that climatic change, even though severe and abrupt, is not an automatic forcing mechanism resulting in directional changes in human societies. Sometimes new solutions are found to subsistence problems, which, rather than encourage overall change to a society, actually act to perpetuate the status quo and work as a stabilizing factor, allowing the subsistence and economic systems to function very much as they did previous to the climatic disruption.

The Early Natufians enjoyed the benefits of the warming climate after the end of the Late Glacial Maximum. They were on the road to developing more settled villages with increasing social complexity. The onset of the Younger Dryas seems to have halted that development, and rather than providing an impetus to agricultural origins, it seems to have impeded developments in that direction. Those developments resumed again only in the PPNA with the return to warmer moister conditions in the Early Holocene, and agriculture truly took hold in the Near East only at the height of the warm wet conditions in the PPNB period.

Early Complex Societies: Climate Change and Collapse of Early Bronze Age Societies

In the previous chapter we began the discussion of relationships between climate change and social responses by examining the impact of environmental stress on hunter-gatherer communities and how this might have lead to major social and economic changes. When examining case studies of hunter-gatherers one could say that the reactions to climate change are more direct and intuitively explicable than in situations involving complex societies. Hunter-gatherers are capable of more immediate responses, sometimes readily abandoning a region that is impacted by adverse environmental change, even when this change is on a short timescale. We can attribute this effect to higher mobility among hunter-gatherers and a greater degree of resilience when compared to societies with more complex social and political organizations. A number of aspects of complex societies hinder them from rapid and flexible change in response to environmental shifts. These include being bound into an agricultural economy, a commitment to sedentary communities, larger population sizes, highly specialized technologies, coercive political control, as well as factors that are more difficult to define such as cultural conservatism and cosmologies incompatible with the necessary economic changes. In attempting to understand the relationship between climate change and early complex societies, many of these social, political, and technological factors must be taken into account, as described in more detail in chapter 1. This chapter examines the case study of an early complex society in the Southern Levant, the Early Bronze II–III polities, which were the first large towns or cities to exist in the region throughout most of the third millennium BC. It shows how this society was impacted by the effects of landscape changes brought on by a major deterioration of the climate and attempts to demonstrate that the ultimate collapse of this society could have been avoided by simple technological changes. For reasons unique to this cultural group, these changes were not adopted.

ENVIRONMENTAL CONDITIONS IN THE THIRD MILLENNIUM BC

The landscape and climatic conditions of the third millennium are discussed in more detail in chapter 5, but a brief review is necessary to emphasize the environmental context of Early Bronze Age societies specific to the Southern Levant before discussing the social and technological context of the period. Isotopic evidence has shown that the third millennium BC was generally a period of greater overall rainfall than that of the present day, with major wet phases peaking at short intervals within the EB I (ca. 3500–3000 BC), EB II (ca. 3000–2700 BC), and EB III (ca. 2700–2200 BC) periods (see the period from 4000 to 7000 BP in figure 5.3 in chapter 5). As pointed out in chapter 5, a major characteristic of the third millennium BC was the marked fluctuation in wet and dry periods. If one were to look only at the average, it would have been a significantly wetter period than that of the present-day dry episode. However, as discussed in chapter 1, subsistence farmers in marginal environments do not respond to overall rainfall averages; they adjust their subsistence strategies to the frequency and intensity of drought in any given year or sequence of years.

Subsistence stress is most difficult to deal with in situations of environmental instability, and farmers as well as hunter-gatherers are best off in periods of stable climate, whether wet or dry (Winterhalder 1990). The impact of instability was particularly acute on preindustrial complex societies. Among these groups it was often the responsibility of elite managers to devise methods for buffering the food supply against periods of unpredictable and unstable weather conditions. If they failed to do so, they would be in danger of losing the support of and control over the subsistence base of food producers.

Although rainfall patterns suffered from a great deal of fluctuation in the third millennium BC, there is much more to the environment than solely the atmospheric conditions. Landscape plays a major role in subsistence economies, and the Early Bronze Age societies of the Southern Levant lived in a landscape that was quite different from the one we see in the region at present. Geomorphological and pollen studies in the area have indicated that the bare hills that we see today were still covered by thick terra rossa soils held in place by oak-pistachio forests and Mediterranean maquis in the north of Israel (M. Zohary 1973; Baruch and Bottema 1999) and lighter loess soils covered by steppic vegetation in the south (M. Zohary 1973; Danin 1983). However, the most important landscape resource lies in the valley bottoms. Today the Southern Levant is characterized by deeply incised stream channels with gravely beds in central and southern Israel, and by underfit streams choked with heavy clay alluvium in northern Israel. At present these streams are dry throughout the summer and flow only during the winter rainy season. In most years when these streams are in flood, they do not overflow their banks, and the water never reaches the surrounding plain. Thus today, all farming within the valley bottoms is primarily rain-fed farming unless crops are artificially irrigated with piped-in water.

The hydrological setting in the third millennium BC was quite different. At that time the streams were not deeply entrenched as they are at present, and rather than a dry summer channel and winter flash-flood regime with flooding restricted to the channels, the streams were characterized by gentle low-energy floods in the winter and regularly overflowed the channel banks and inundated the floodplains with freshly deposited silts (Goldberg and Bar-Yosef 1982; Goldberg and Rosen 1987; Goldberg 1994; A. Rosen 1989, 1991, 1997b). There is also evidence to suggest that even in the summer dry months the streams were more perennial with a low-energy low-volume flow (A. Rosen 1986b, 1991).

The evidence from Tel 'Erani provides a good example of this landscape change (see figure 7.1). Here, geomorphological reconnaissance revealed the existence of terraces dated to the Early Bronze Age (A. Rosen 1991). The terraces are composed of sediments that indicate several facies of contemporaneous deposition. One unit is a gravel deposit representing the stream channel, but more importantly, there were at least 3 m of fine-grained sediment that was composed of finely laminated silts and clays. These sediments tell a story about the depositional environment at the time. Specifically they represent the deposits of overbank flooding from a slowly rising steady flow with gentle to medium energy. These are the levee and backswamp deposits from a regularly flooding stream (A. Rosen 1991). Similar deposits are found throughout the Southern Levant. In some of the sediment sections dating from this period there are remains of heavy dark clayey deposits representing

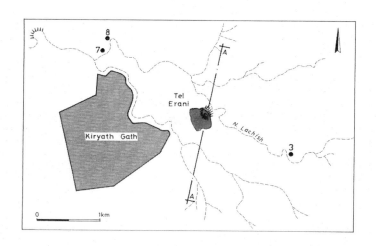

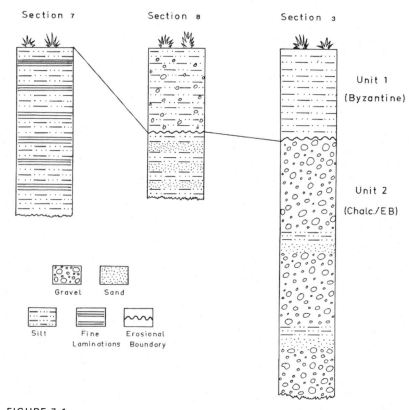

FIGURE 7.1

Map of the Vicinity of Tel 'Erani Showing Locations of Wadi Sections with Chalcolithic/EB Alluvium, and Transect across the Ancient Floodplain (A. Rosen 1991).

seasonal marshy deposits. These are seen in the area around Lachish (A. Rosen 1986c), Har Tuv in the Shephela (A. Rosen, unpublished manuscript), and as far north as the plains around Megiddo (A. Rosen 2006).

A final note on the landscape and environment of the Southern Levant in the third millennium BC comes from pollen data. As pointed out by Baruch (1986, 1990), pollen data become less reliable as a climatic proxy during periods when agriculture is intensive and when humans begin to impact the landscape in a significant manner. Although geomorphological evidence from northern Israel has shown that there is very little soil erosion and therefore only limited deforestation during the Early Bronze Age (A. Rosen 2006), the increase in the pollen peaks for *Olea* in the Hula and Kinneret cores suggest that the production of olives for oil manufacture had impacted the vegetation composition of the forest (Baruch 1986, 1990).

SOCIAL DEVELOPMENT DURING THE EARLY BRONZE AGE
The Early Bronze Age in the Southern Levant is the period in which the first large towns and cities emerged. Although a few major centers appeared in parts of the landscape in EB I, notably Tel 'Erani in southern Israel and Tel Megiddo in the north, it wasn't until EB II (3000–2700 BC) that the large towns began to dominate the political landscape of the region (Joffe 1993). The development of this townscape began with the movement of small villages from the floodplains near the drainage channels to low hills and rises. These EB II towns were the foundations of what later became major urban mounds or *tell* sites after thousands of years of occupation (A. Rosen 1986a).

For the first time in this part of the Near East populations had grown to large proportions and organized into an urban landscape. Some scholars suggest that the population in the Southern Levant rose from approximately 2400 in the EB I to around 35,000–40,000 in EB II–III (Gophna and Portugali 1988; Finkelstein and Gophna 1993). Settlement also shifted to a greater emphasis on the strip of land along the eastern edge of the coastal plain, in the hilly margins (known locally as the Shephela) bordering the coastal plain on the west and the hills to the east (see chapter 3; figure 7.2). This region had an obvious economic advantage for controlling produce from the cereal-producing coastal plain and the olive-producing Shephela Hills.

The period is also characterized by increasing social complexity. This complexity is reflected in the emergence of monumental architecture including massive fortifications and temple complexes at sites such as Tel Yarmouth (de

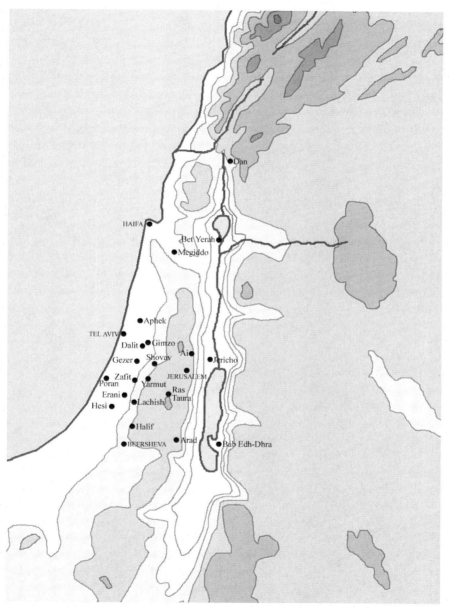

FIGURE 7.2
Map of Major Early Bronze Age Sites (Prepared by D. Beatty).

Miroschedji 1988), Megiddo (Kempinski 1989), and Ai (Callaway 1980), as well as granaries such as that at Beit Yerah (Esse 1991). These architectural features accompanied the appearance of a ruling elite (Joffe 1993), significant trade with Egypt in the EB II period (Stager 1985; Esse 1991), and craft specialization (S. Rosen 1997).

There is evidence that the ruling elite maintained control over both the produce of the cereal-growing region of the coastal plain and the production of olive oil and wine in the hilly regions (A. Rosen 1995, 1997a). This comes from the large granaries excavated at Tel Beit Yerah at the southern end of Lake Kinneret (see figure 7.3) (Esse 1991). Although this is a unique find in the region to date, it is likely that there were other such granaries that have not yet been encountered at other sites. Other types of mass storage have been found at Tel Yarmouth in the southern Shephela hills. Here, excavators found major storerooms within an elaborate palace complex consisting of compartments full of large storage vessels (de Miroschedji 1988). One room in the temple was filled with ash rich in the phytoliths of wheat (A. Rosen, personal

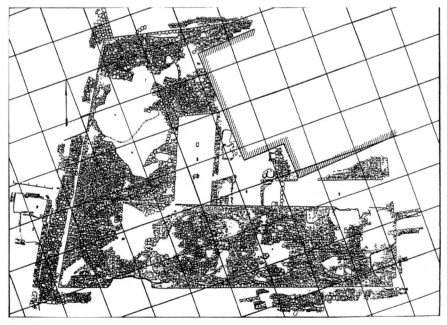

FIGURE 7.3
Early Bronze Age Granary at Beit Yerah (After Maisler, Stekelis, and Avi-Yonah 1952).

observation). Storage of mass quantities of goods implies a large-scale appropriation of public produce.

Given the above, one might wonder about the coercive power of a newly emerged elite class for extracting a grain or oil tax, or tribute from the subsistence farmers within the catchment of a given site territory. At the same time as the emergence of these urban settlements there was an increasing investment in the building and maintenance of temple complexes. This phenomenon is closely related to the rise of the elite classes, the increasing importance of international trade, and the growing population. It is reasonable to assume that the temple cults played a key role in the extraction of goods and services from the population that supplied the subsistence base of the city. However, this was probably not all a one-way street. The temples would have also provided essential services as perceived by this population in the form of guaranteeing the fertility of the land and ensuring the regularity of the rainfall, and when drought years did grip the countryside, they probably also supplied emergency rations in the form of redistributed grain from the storerooms.

Another key element in the picture of Early Bronze II–III economies was the role of trade in olive oil and wine (Stager 1985; Esse 1991; Joffe 1993). Trade in these key Mediterranean crops formed the foundation of the wealth and power of the rulers, administrators, and temple officials (Joffe 1993). This trade reached a peak in the EB II, as we know from evidence derived from Egyptian sites where there have been many finds of storage vessels (Abydos ware) originating in the Levant (Ben-Tor 1986). These vessels were transported as containers for wine and olive oil. The international trade with Egypt ceased in EB III, but Joffe (1993) suggests that the production of olive oil and wine as cash crops continued due to the demands of the internal Southern Levantine markets. Revenue from the production of these crops continued to flow into the coffers of the managing elite due to these increasingly important local markets.

By the end of the third millennium BC, there was a total collapse of society with general abandonment of almost all the urban centers. Population decreased dramatically from more than 14,000 (Gophna and Portugali 1988), or up to a possible 40,000 in southern Canaan alone (Finkelstein and Gophna 1993; Gophna 1998), in the EB III down, to an estimated 1800 in EB IV (see figure 7.4) (Gophna and Portugali 1988). For the most part the remaining population dispersed and settled into small farming villages and/or returned to an economy of seminomadic pastoral nomadism (Dever 1998). This is an

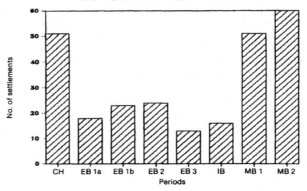

NUMBER OF SETTLEMENTS

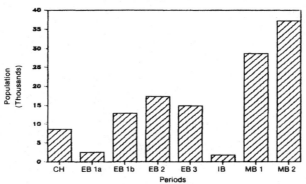

CALCULATED POPULATION

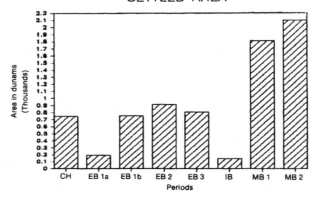

SETTLED AREA

FIGURE 7.4
Population Shifts from EB I through EB IV (From Gophna and Portugali 1988).

archaeological period known alternatively in the literature as EB IV, MB I, or EB–MB Intermediate period (from 2200 to 2000 BC). I refer to this period as EB IV to emphasize the point that this is very much a continuation of the same population and traditions of the Early Bronze Age.

LANDSCAPE CHARACTERISTICS AND THE AGRICULTURAL ECONOMY OF THE THIRD MILLENNIUM BC

To briefly summarize the lay of the land during the third millennium BC with a view to the agricultural economy of the period, we can divide the landscape of the Southern Levant into the *hilly regions* and the *lowlands*. The hilly regions include the Judean and Samarian hills, which form a spine along the eastern margins of the region, and the Galilee Hills further north. The lowlands consist of the broad alluvial valleys of what today is northern Israel and the coastal plain of central Israel.

Hilly Regions

The hilly regions of the Galilee, Judea, and Samaria receive higher rainfall than other parts of the country. Today's annual average rainfall for these areas is approximately 1000–500 mm (Orni and Efrat 1980). Although the highlands were suitable for small-scale dry farming in the narrow valleys, intensive cultivation of subsistence crops under the agricultural technology of the third millennium BC Early Bronze Age towns was unlikely. This was due primarily to the steep slopes and dense forests. Although the forests could have been cleared, without some form of terracing the rich agricultural soils would have been rapidly lost due to soil erosion and solifluction after a short number of years of cultivation. Agricultural terracing does not make an appearance in the area until much later, specifically in the Iron Age of the first millennium BC (Borowski 1987). Thus, to exploit this rich region agriculturally, the most lucrative strategy would have been to utilize the area primarily for tree crops, namely for olives and viticulture (Finkelstein and Gophna 1993:11–12).

There has been much discussion in the literature about the exploitation of olives and grapes for oil and wine, and their role in propelling the Levant into a world economic system based on international trade (Stager 1985; Esse 1991; Joffe 1993). These crops played a major role in turning the region from a backwater of small village farmers into a more complex network of competing polities, organized into independent towns with large populations. However, in addition to these broader implications for the development of society and bringing this area onto the world stage, a more internal economic view would be that the use of the hilly areas for olive and grape production was a way of optimizing production to insure the subsistence base of local populations as well. In years with high cereals yields, revenues from olives and grapes could be turned into luxury products and other goods such as textiles, metal objects, slaves, herds, and so on. In years with low yields of cereals, both the output from the oil and wine production and the stored luxury goods could be turned into food by the use of these products as a currency for purchasing grain from other regions. Such strategies are well known in historical and modern cases of nonindustrial farming societies (Colson 1979; Halstead and O'Shea 1989b).

Archaeobotanical evidence from sites such as Tel Yarmouth (Lipschitz 1989), Lachish (Helbaek 1958), and Jericho (Hopf 1983) suggest that olive and grape production was indeed a major part of the Early Bronze Age economy in addition to the growing of cereals and other subsistence crops. The social implications of this cash-crop production are discussed in more detail below.

Coastal Plain and Northern Alluvial Valleys

These areas are characterized by heavy clay-rich soils, often high in organic matter in the north and center of the region, and lighter more loamy soils in the southern coastal plain and northern Negev Desert (Ravikovitch 1969). As discussed above, the alluvial valleys and coastal plain in the third millennium BC enjoyed a much different hydrological regime from that of the present day. Today, many of the streams of this area are deeply entrenched and flood their banks only in the rainiest years. When they do flood their banks, only small strips of land on either side of the channel are impacted by the floodwater. Therefore, under the present-day hydrological regime floodwater farming would not be a viable option.

The geomorphological evidence for seasonally inundated floodplains during the fourth and third millennia BC has important implications for the way in which farming was conducted during the Chalcolithic period and Early Bronze Age. Unlike the situation today and throughout most of the Late Holocene, the opportunity existed for floodwater farming within the lowland valleys of the Southern Levant. This farming strategy would have provided a significant buffer against the frequent drought years experienced in the region, even during this Middle Holocene episode of increased overall rainfall.

The advantages of this farming strategy were not lost on the Early Bronze Age farmers. Evidence of EB period floodwater farming comes from analyses of phytolith remains from Early Bronze Age sites, including remains from the palace storage facilities at Tel Yarmouth, Israel. Phytoliths are microfossils of plant cells that form in stems and floral parts during the life of the individual plant. They are most abundant in grasses (including cereals), sedges, reeds, and palms. When the plant tissue disintegrates, the phytoliths remain behind in the archaeological sediment and are a testimony to the type of plant that once existed at a particular location on the site (A. Rosen 1999). Rosen and Weiner (1994) have shown that in semiarid contexts, cereals grown in wet alluvial fields produce many more joined silicified cells (or silica skeletons) than those grown in well-drained soils. Thus silica skeletons can be used to distinguish between plants grown in irrigated fields (including floodwater irrigation) and those grown under dry-farming conditions.

At a number of Early Bronze Age sites, including Tel Yarmouth, the phytoliths are characterized by unusually large silica skeletons consisting of hundreds of silicified plant cells per phytolith (A. Rosen, personal observation). These suggest that wheat was cultivated in heavy moist soils. Since we have no evidence for irrigation canals or dam structures dating to the Early Bronze Age, it is most likely that these cereals were grown in wet fields using a form of floodwater farming or simple basin irrigation.

The use of floodwater farming was probably not exclusive of other types of farming practices. It is likely that dry farming was conducted extensively as well. However, yields from dry farming would have been more subject to great fluctuations due to the unpredictability of rainfall from year to year. It is very likely that the key to supporting large populations in this region, with a low level of agricultural technology, would have been the use of floodwater farming to increase the predictability of the yields from year to year and thus buffer against periodic droughts. This would have guaranteed a stable source of grain even in dry years with their reduced yields from dry-farmed localities.

It is significant that EB II settlement is characterized by a movement off the floodplain onto low hills, Pleistocene stream terraces, or more mountainous areas. First, this had the practical benefit of putting the town out of danger from flooding. Second, it freed up more agricultural land. And finally, there might have also been a defensive advantage to building the towns at a height above the valley bottoms.

MODEL OF EARLY BRONZE AGE FARMING STRATEGIES

Given the above brief reconstruction of the third millennium BC landscape in the Southern Levant, and the socioeconomic and political background of the period, we can propose a model for agricultural strategies undertaken in the Early Bronze Age. Surveys of Early Bronze Age sites have shown that most of the major centers were situated along the western boundary of the Shephela, bordering the mountains and the coastal plain (Gophna and Portugali 1988). It is likely that this was economically strategic because it allowed the urban centers to control production in both the hilly areas as well as the lowlands. There was thus a topographic as well as economic duality to production. The horticultural products, primarily olives, olive oil, and wine, were most likely produced in the hilly regions of the Shephela. This was ecologically advantageous because it would have preserved the hillslope soils better than cereal farming on the slopes, and it could take advantage of the higher rainfall in these areas. These products were cash crops, and as such the production, trade, and revenue were controlled by the elite (Joffe 1993). This was the income that allowed the acquisition of luxury goods, the building of urban fortifications and temple complexes, and the support of craft specialists.

As a cash crop managed by the elite segments of society, the decision-making processes and risk taking in times of environmental stress would be different from those undertaken by subsistence farmers. These differences are outlined in a general way in chapter 1. Rather than a strategy of risk minimization as employed by subsistence farmers, a more common cash-cropping policy would be to maximize the yield even though the risks of losing the crop might be higher. Such a tactic would lead to intensive monocropping. In the case of olive oil production this would mean grove after grove of olive trees rather than interspersing these with other types of trees. Monocropping has the potential of higher output, but with the risks of increasing the spread of plant diseases and the possibility of losing the entire crop to environmental conditions unfavorable to olive trees in particular. This suggestion finds some support in the pollen diagrams from Lakes Kinneret, Hula, and Birket Ram (Baruch 1986; Baruch and Bottema 1999; Schwab et al. 2004), which show increases in olive pollen at the expense of other tree species during the third millennium BC. In times of short-term drought, yields from cash crops would drop. In extended periods of drought, one maximizing strategy would be to increase the numbers of orchards to maintain expected overall production of the crop. This increase would come at the expense of time, energy, and land resources that might have increased production of subsistence goods.

A contrasting strategy would have been employed by subsistence farmers operating primarily in the lowland valleys and coastal plain. We know from many ethnographic analogies that subsistence farmers in semiarid areas utilize methods of risk minimization, rather than implementing policies to maximize yields as basic modes of operation (see chapter 1). To reiterate, subsistence farmers in marginal farming regions employ a system of mixed cropping, thus insuring a yield of edible foodstuff, even if the preferred crop is adversely impacted by poor rainfall for one or several years. In the case of Levantine farmers, the preferred crop has traditionally been wheat, with barley raised as a less desirable backup commodity. Wheat is also more vulnerable to drought than barley (Renfrew 1973). Although there have been precious few systematic studies of paleobotanical remains from Early Bronze Age II–III sites, it is clear that both emmer wheat and barley form a significant part of Early Bronze Age archaeobotanical assemblages (Helbaek 1958; Hopf 1983; Liphschitz 1989). Another critically important aspect of making a living in marginal farming regions is the ability to manage surplus with an eye toward drought years even in times of plenty. Subsistence farmers have sophisticated storage schemes to overcome droughts of one and sometimes two years. To buffer themselves against droughts that last longer than two years it is necessary to convert perishable foodstuffs into nonperishable goods that can later be sold or traded for food in times of need. In the case of the Early Bronze Age farmers, such goods might include copper items and perhaps fabrics (Ilan and Sebbane 1989).

This system of buffering against periods of extended drought depends very much on the ability of farmers to have full control over their surplus production. In the case of the Early Bronze Age polities, the large communal storehouse at Beit Yerah (Esse 1991) suggests that in fact the state controlled at least a portion of surplus produced by subsistence farmers. This *tax* or *donation* of foodstuffs to a centralized authority could have been both coercive and voluntary. It was voluntary in the sense that it might have been perceived as a necessary contribution to the support of the temple cults that were responsible for maintaining fertility of the land as well as the bringing of rain (R. Amiran 1972; Ben-Tor 1992:116–118). There is some evidence for the role of temple cults in this capacity from other regions of the Near East including Mesopotamia (Kramer 1966; Jacobsen

1970). For the long term, the diminished control over surplus would have inhibited farmers from reacting to droughts on an individual basis, reduced their resilience, and increased their reliance on the state for help in times of famine. This reliance on the state for redistribution of grain would have increased the coercive power of the elite managers over the general population. During times of extended droughts, if the state was unable to supply a growing demand for food, then subsistence farmers would have no recourse to alleviate their stress, and a disastrous situation would occur.

END OF THE EARLY BRONZE AGE III

The abandonment of sites across the Southern Levant was well underway by 2300 BC (Seger 1989). Some EB III sites such as Tel Halif in the northern Negev show signs of destruction and conflagration. Other sites such as Tel Hesi, also in the northern Negev, simply seem to be abandoned at the end of the Early Bronze III, with occasional occurrences of squatters' houses marking the final occupation of the settlement. The demographic pattern of the succeeding EB IV period shows an increase in the number of sites, but these sites are very small in size and reflect a significant overall decrease in population numbers (Gophna and Portugali 1988; Gophna 1998). In the highlands, all sites were deserted during the EB IV (Finkelstein and Gophna 1993), indicating a breakdown in the horticultural infrastructure that had produced abundant revenues from the production of olive oil in the preceding period. These abandonments indicate there was a momentous collapse of Early Bronze Age civilization in the Southern Levant, a phenomenon that has been mirrored in many other locations throughout the Near East.

The following period of the EB IV (2200–2000 BC) is marked by a number of small village sites and pastoral encampments. The change in subsistence and economy between the EB III and the EB IV was a profound one. There was no longer a superstructure of elite control over production in the subsistence sphere, and there was a collapse of the cash-crop system of oil and wine production. The remaining population of the region returned to small social groupings in nonhierarchically arranged villages and nomadic pastoral camps (Dever 1998). There was a dominance of small-scale agro-pastoral economies. Many of the sites from this period are located close to springs and perennial water sources. The exploitation of available springs may explain the renewed settlement in the Negev Desert by pastoral nomadic groups, after a virtual abandonment of this region during the EB III (S. Rosen 1987).

However, this general return to smaller settlements was not universal. In Transjordan, two large urban centers continued to exist and thrive in spite of the loss of population in the surrounding area. These sites are Iskander (Richard 1990) and Iktanu (Prag 1989). These two settlements are unique in that they show no signs of abandonment at a time when urban collapse had been taking place all around them for decades. The question of why some towns suffered collapse of the social order and a complete abandonment while others appear to have been unaffected by events at the end of the third millennium is of critical importance for understanding the relationship between preindustrial states and abrupt environmental change. This issue is discussed in more detail below.

ENVIRONMENTAL CHANGE AT THE END OF THE THIRD MILLENNIUM BC
The evidence for environmental change at the end of the third millennium BC is reviewed in chapter 5. Although any one line of evidence might be suspect, the various sources of environmental information point to a decrease in overall rainfall, a regression of the Levantine forests, and a change in the stream hydrology. These effects not only impacted the Southern Levant but were widespread across northern Syria (H. Weiss et al. 1993; Courty 1994), and Turkey (Lemcke and Sturm 1997; A. Rosen 1997c, 1998). Given these changes there would have certainly been an impact on the societies that existed in the region at this time. However, this impact was not in itself the sole cause of widespread collapse. If this were the case, it would be difficult to understand why some towns continued to exist while others around them failed.

To comprehend this phenomenon and the relationship between climatic factors and the archaeological evidence, we have to take into consideration the unique social, economic, and political aspects that operated in these Early Bronze Age societies. This becomes evident when we consider that environmental degradation at the end of the third millennium BC was a spike of bad climate heralding a longer-term directional change, not simply a short-lived severe drought as suggested by Harvey Weiss et al. (1993). The climate became drier at the end of the third millennium BC and remained dry with the exception of a few spikes of moister climate, generally until the present day (Goodfriend 1988, 1991; Bar-Matthews, Ayalon, and Kaufman 1998). The significance of this for archaeological interpretation is that throughout the succeeding millennia, societies during the Middle Bronze, Late Bronze, and Iron Ages had to contend with the feeding of cities and large populations under the same dry conditions that were in part responsible for the downfall of the Early Bronze Age societies. This was accomplished by improved technologies for agriculture and water storage, as well as the political and economic infrastructure for storing and redistributing food in times of hardship.

The above-mentioned climatic factors point to a general decrease in rainfall in the Near East after the end of the EB III. The impact of this on subsistence and cash-crop farming in decreasing yields is obvious. Less obvious, though, is the impact on the hydrological systems of the region and their role in removing what may have been an important buffer against the effect of periodic drought. Throughout most of the Early Bronze Age the streams that drained the lowland valleys of the Levant were alluviating and flooding their banks during the winter rainy season, allowing the practice of floodwater farming. At the end of the third millennium BC, the decrease in rainfall led to a significant change in stream hydrology. The effect of this was a lowering of base flow in the streams, causing many of them to change regimes from perennial to ephemeral. The overall effect was for the streams to begin downcutting their channels, resulting in steep banks, a cessation of annual overbank flooding, and an abandonment of their formerly active floodplains. The floodplains were left high and dry, thus removing a key agricultural resource, namely the heavy moist soils that had been renewed on a yearly basis. Without this essential buffer, the population was left in a precarious position with respect to the annual food supply. Surpluses could no longer be guaranteed, which would have had disastrous effects on the large populations living in the region at the time.

MODEL FOR THE ROLE OF ENVIRONMENTAL CHANGE IN THE COLLAPSE OF EB SOCIETIES

At the end of the third millennium BC there was a dramatic environmental change. This was accompanied by a demise of Early Bronze Age III civilization in the Levant as well as throughout much of the Near East, which witnessed a profound decrease in population and abandonment of towns and cities. Although both of these events, climatic and social, are equally catastrophic, we must be cautious about drawing a one-to-one parallel between climatic cause and social effect. I illustrate this point by calling attention to the Southern Levantine towns of Iktanu (Prag 1989) and Iskander (Richard 1990) located on the eastern side of the Jordan Valley. As indicated above, these towns were not abandoned at the end of the third millennium BC. Although towns and villages were being abandoned all around them, these sites seem to have been relatively unchanged by either the climatic or demographic events in the region. The evidence for this is the continuation of settlement levels throughout the period of the EB IV, including the maintenance of a city wall. This latter point is taken as evidence for the continued dominance of city elite and ruling classes (Richard 1990). It is possible that part of the population of these towns included refugees from other sites, which might partly account for the appearance of a stable population. However, as yet we have no archaeological evidence for this.

One factor that may have accounted at least in part for the apparent stability of the two sites of Iktanu and Iskander is the fact that both are located on what even today are perennial streams fed by permanent springs. If the settlements of Cisjordan (west of the Jordan River) were stressed by the lack of natural irrigation water after hydrological changes in rain-fed streams, then the occupants of these Transjordan (east of the Jordan River) sites might have avoided that problem by their more fortunate location. Evidence that surface water availability might have been an important factor in the continued existence of a city at the end of the Early Bronze Age comes also from central Syria. The city of Mari, located on the central Euphrates in the Syrian Desert was another example of a large population center that exhibited little evidence of any impact from decreased rainfall at the end of the third millennium BC (Finet 1985; Geyer 1985; Sanlaville 1985). However, this was clearly not the only factor in the successful adaptation to the new climatic regime, since other sites in the Jordan Valley suffered urban collapse in spite of their proximity to perennial water sources. Such sites include Jericho (Kenyon 1960) and Bab edh-Dhra (Rast and Schaub 1981).

Decreasing rainfall and diminishing surface water placed the burden of feeding relatively large populations firmly in the lap of the ruling elite, who were entrusted with the responsibility of propitiating the gods and thus ensuring adequate rainfall, soil fertility, and ultimately food supplies. The local subsistence farmers were the key food producers in this early complex society; however, the farmers were required to entrust their surplus produce to the central control of the state. In agreeing to do this they benefited from the protection of the state, received the blessings of the religious establishment, and were probably the beneficiaries of redistributed goods in years when their crops failed. The disadvantage of this relationship in the long term was that the subsistence farmers had to relinquish a large measure of their resilience by giving up control over their surplus goods. With the diminishing ability of the elite managers to feed growing numbers of failed farmers, their coercive power over the population would also have declined.

A simplistic view of the economic situation at the end of the third millennium BC might assume that this catastrophic situation was an unavoidable consequence of the clearly adverse climatic shift. However, the environmental changes were only part of the problem. Given the fact that after the Early Bronze Age, large settlements were founded in this region from the second millennium Middle Bronze Age onward, and many of these large towns and cities had to cope with the same adverse climatic conditions as those that faced populations at the close of the third millennium BC, we must also look for a social and/or political reason for the widespread failure and resulting abandonments at the end of the EB III period.

With this fundamental stress on the agricultural sector of society brought on by environmental degradation, we must question the responses of the elite managers to the reduced yields of subsistence crops, the resulting decrease in revenues, and diminishing popular support. They seem to have taken no action in the technological, social, or economic realms to improve the situation. One possible solution might have been the introduction of canal irrigation. Canal irrigation must have been known to this society, which had had numerous trade and cultural contacts with both Egypt and Mesopotamia. The irrigation system need not have been an elaborate structure of canals and dikes. A simple mud and rock dam near a stream or upper reaches of a water course could have redirected water to flow into fields rather than run off through the entrenched channels downstream. This would not have taken a major regulation of labor and could have been accomplished on a village level of organization.

One possible reason for the failure of the agricultural sector to adapt to the new hydrological regime might be that the response time of the elite managers was too slow. If the social and political infrastructure was used for the redistribution of food during the course of normal droughts, and these occurred at regular intervals, then the impact of long-term drought might have been misjudged until it was too late to activate a substantial change in agricultural practice or technology. In other words, the established system of provisioning the population was well adapted to the old climatic regime, and the better such systems function, the harder they are to change.

Another possibility for the slowed response time has to do with the phenomenon in which elite landowners benefit from drought, floods, and other catastrophic events that impact subsistence farmers. In the short term, this situation leads to a greater availability of low-cost labor and cheap land prices, because many farmers are forced to sell their labor and land to survive. With the expectation of increased profits during such times of economic stress, the elite managers may have been slow to respond to the declining productivity of the land, not realizing the seriousness of the problem until it was too late to take action for recovery.

Furthermore, we should also consider how perceptions of weather, climate, and environmental change might have influenced the behavior of the Early Bronze Age peoples in the Southern Levant. These perceptions and the resulting interpretations of how to deal with climatic change might have been quite different from those in Western societies. Today in the Western world we think very much in terms of scientific and technological solutions to our own problems of climatic change, specifically global warming. Although it is difficult to get a glimpse into the minds of members of ancient Near Eastern societies, we know from Mesopotamian texts that climate, perceived on the short timescale as weather, was believed to be controlled by deities who lived in a very complex cosmological framework. These storm gods controlled both rainfall and wind. In the *Ba'al Cycle*, Ugaritic texts dated to the mid-fourteenth century BC, the storm god Ba'al controlled the frequency and strength of rainstorms, and the method of communicating with this god was through the "window" of temples dedicated to his service. The intermediaries to the gods were the kings and priests. These ruling classes wielded great power by their perceived ability to control rainfall and fertility of fields (Smith 1994).

Ruth Amiran (1972) and others (Ben-Tor 1992:116–118) have suggested that similar religious beliefs existed in the Southern Levant. This is based partially on similarities between images on items. For example, Amiran (1972) argues that the representation of the anthropomorphized figure of a dying wheat sheath on a stela from EB Arad in the Negev Desert has striking affinities to the Mesopotamian harvest god, Tammuz. If we accept that there was a strong belief that the gods controlled weather patterns and particularly rainfall, then we can suggest that the logical response to the lack of rain would be first to appeal to the gods to convince them to bring back normal rainfall. In the framework of Near Eastern religious practices, these first appeals might take the form of sacrifices, but in more extreme cases, energy and resources could have been diverted to major public works projects such as temple building in honor of the deities (A. Rosen 1995).

If we look to historical analogies for support of this suggestion, we have the striking example of that very response on the part of the Greenland Norse, based on work by McGovern (1994), cited in chapter 1 of this volume. When the Greenland Norse found themselves under severe environmental stress from the onset of the Little Ice Age in the fifteenth century AD, rather than change subsistence strategies to the adaptive pattern of their neighbors the Inuit, they concentrated their efforts on building a massive cathedral and trials of witches and other unbelievers.

Temple building was an essential activity of city life in the Southern Levant from the EB II through the EB III, and most major population centers boasted temple complexes. This activity was stepped up throughout the EB III with larger and more complex temples dominating the townscapes, as was the case with the impressive temple complex at Megiddo (Kempinski 1989). The above hypothesis suggests there was increased temple building with increasing uncertainty in rainfall patterns. If this was the case, then it is possible that as climatic conditions worsened throughout the EB III, the priests and managers put more effort into the temple cults in an attempt to gain a greater hold on the population by using their perception that the priests and rulers controlled the gods of rain. On the more practical side, the population would have also depended upon the ability of the rulers to redistribute grain from public stores in times of intense need.

To summarize this position, some key factors for understanding Early Bronze Age adaptations and the impact of climatic change on the society rest on three prime elements. One of these is the way in which the society was organized, the second is the importance of a temple fertility cult, and the third is the relatively low level of agricultural technology in this period. The EB II and III populations of the Southern Levant had adapted themselves technologically to a climate of periodic drought by buffering yields with simple floodwater-farming techniques, and socially by the public storage and redistribution of grain stores during times of drought. The authority to exact a grain tax and to allocate the surplus was given to the priests and elite rulers based on the belief that they were the intermediaries to the gods and thus guardians of rain, fertility, and abundance of food. With the changing climatic regime at the end of the third millennium BC, the resulting shift in stream hydrology to one of alluvial downcutting removed the advantage of floodwater farming and led to the breakdown in simple floodwater irrigation systems. This seriously reduced yields of subsistence crops and led to increased numbers of famine years and general social stress. The elite managers might have been slow to respond with a technological solution to the problems, such as better water capture and canal irrigation systems. This critical delay might have led to widespread famine, disease, death, and dispersal of the population.

The exceptions to this scenario at the Transjordan sites of Iktanu and Iskander were possibly due to the elite rulers being able to avert impending disaster through the implementation of improved agricultural technology and/or social reforms. This is something that can be tested in future work at these sites perhaps by looking for evidence of check dams or other irrigation schemes.

This chapter focuses on a study of an early complex society located in a marginal farming zone and its responses to climatic and natural environmental degradation. This society had a low level of agricultural technology, but a sophisticated social organization that had devised a successful adaptation to supporting large populations in this semiarid zone. Although the climatic change was a significant stress to the social and economic system, it was not the sole reason for the demise of Early Bronze Age society in the Southern Levant. It is in the exploration of these extenuating details that we are able to acquire much insight into the way in which societies responded in the past. These insights might also help us to understand the ways in which modern societies still react to episodes of climatic deterioration.

8

Empires in the Desert: Political Ecology of Ancient Empires

Earlier chapters discuss the role of climate as one factor that contributed to social, political, and economic change in hunter-gatherer and early complex societies. This chapter examines the impact of climate change on one of the most complex organizations of human societies, the empire. In the previous two types of sociopolitical organizations, the environment and shifting climatic conditions had a very direct impact on the immediate survival of the society. Under severe conditions of climatic stress, the test of whether societies survive or not can be found in the resilience and viability of their strategies for adaptation to the sometimes abrupt onset of climatic shifts.

In antiquity, empires are arguably the types of societies that most closely resemble our own, with similar global political and economic infrastructures connecting nation-states worldwide. As with other societies, it is difficult to isolate climatic factors and human responses without considering politics, economies, and ideologies. The examination of climate change and its impact on empires in antiquity not only gives us the advantage of observing the influence of climate change on complex societies across a broad spatial extent, but also provides the benefit of time depth.

Some researchers have argued that in spite of their complex economic interdependencies, even empires cannot withstand the pressures on agricultural systems that are brought on by climate change. This can be a direct result of decreased yields leading to immediate food shortages, famine, disease, and social unrest (Brown 1995; Garnsey 1988; Lamb 1995), or the indirect effect of economic stress resulting in invasions by neighboring societies. In 1907 (in *The Pulse of Asia*), Ellsworth Huntington proposed that the fall of the Roman Empire could ultimately be traced to severe climatic shifts in central Asia that led to the drying up of pasturelands, thus triggering a series of westward migrations on the part of the nomadic peoples living in the region. The result of this was the displacement of peoples as far west as the Mediterranean and the attacks on Rome by migrating tribes of "barbarians." More recent scholars have adopted this concept, applying it not only to ancient civilizations but to our own as well (Brown 1995; Issar and Zohar 2004; Lamb 1995).

However, the impact of climate on empires is neither simple nor direct. In the case of empires, the support networks are broader, and the social and economic infrastructures are geared for absorbing a great deal of stress to the system. These more complex and sophisticated social structures, and their ties with world systems and markets sometimes stretching over continents and across seas, means that no one segment of an empire is an isolated entity that must suffer the burden of environmental degradation in a vacuum. Given this situation and these conditions, what then is the effect of climatic change on empires, how do they respond, and can climatic shifts be the ultimate cause of the fall of these sophisticated forms of social organization?

European travelers to southern Palestine in the nineteenth and early twentieth centuries, such as Robinson (1837), Palmer (1871), and Wooley and Lawrence (1914) in the years before World War I, noted with awe the ruins of six large towns in the midst of the Negev Desert. The largest of these were Elusa (Halutza) in the northern Negev, Rehobot (Rehovot Ba-Negev), Nessana (Nitzana), close to the modern Israeli-Egyptian border, Soubeita (Shivta) in the central Negev, Oboda (Avdat), near the modern community of Sede Boker, and Mampsis (Mamshit) in the eastern Negev (see figure 8.1). These towns supported large populations during the Byzantine period and represented wealthy communities that boasted monumental architecture and massive churches.

There were also numerous small farmsteads and agricultural installations surrounding the large urban centers. These have been systematically surveyed only in recent years (see for example Baumgarten 2004; R. Cohen 1981, 1985; Lender 1990). The archaeological remains of intensive farming systems in the hinterland surrounding these urban centers testify to a well-developed agricultural system in the desert, far from the agricultural zones of the present day. Early researchers were greatly impressed by the vast extent of the agricultural

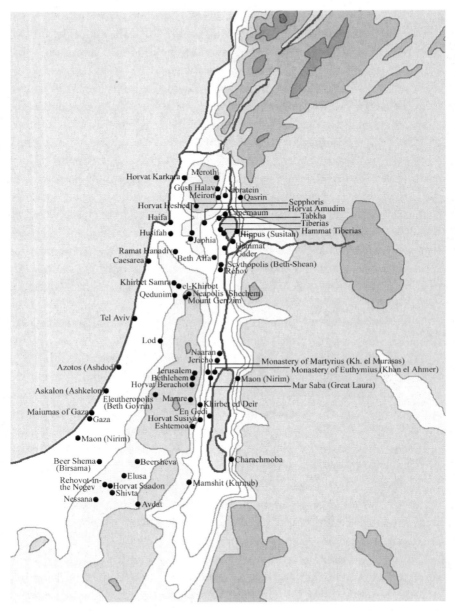

FIGURE 8.1
Map of Byzantine Period Settlements in the Southern Levant (Prepared by D. Beatty).

array of check dams, runoff systems, and numerous field markers found in the vicinity of these abandoned cities, although the initial excavations concentrated on the architecture of the urban centers themselves (Colt 1962; Negev 1988a, 1988b). Excavations of the towns revealed numerous large wine presses, suggesting the processing of locally cultivated grapes and the production of wine at an industrial scale (Mazor 1981).

The discovery of these major towns in the midst of this arid landscape captured the imagination of travelers to the region and led to much speculation about how the populations of these urban centers could have supported themselves. The possibility that the climate was more favorable for agriculture in Roman-Byzantine times was immediately suggested (see the discussion in Rubin 1989). This lead to a great debate between those such as Palmer (1871) and Woolley and Lawrence (1914), who claimed the collapse of the desert civilization was the result of historical and political processes, and those such as Huntington (1911), who believed that the climate was significantly more favorable for agriculture in the Negev during Roman-Byzantine times than at present. This debate became infused with political content when the British Foreign Office adopted Huntington's perspective that semiarid zones were a barrier to civilization and settlement to justify limiting Jewish immigration into Palestine during the Mandate Period. In response to this, proponents of Zionism at the time claimed that desertification in Palestine had less to do with climate and more to do with poor administration and lack of technology (Troen 1989; S. Rosen 2000). The politics of desertification and human responses to climate change continue to play a significant role as concerns over global warming reach critical importance. We can find many parallels to modern circumstances in a study of incursions of complex societies into desert regions. An understanding of why Roman-Byzantine populations chose to exploit desertic regions and how they dealt with the risks of aridification may help us understand some of the responses of modern societies to the problem of global warming in semiarid localities.

As indicated by the review of climatic proxies for this period in chapter 5, summarized below, we still do not have fine enough resolution in the dates of climatic data to be able to clearly mesh climatic fluctuations with precise historical events in the Roman-Byzantine period. However, one important point that must be stressed about the climate of the Roman-Byzantine period is that although there were certainly numerous rainfall fluctuations throughout the Roman period and also within the Byzantine period of the Southern Levant, there was no singular very wet or very dry phase that corresponded to the whole of these time periods. These vacillations were relatively minor changes on the overall scale of Holocene climatic history, although they might have been enough to impact farming in this semiarid environment. Any attempt to simplify the climatic picture by "wiggle-matching" the wet and dry periods with the rise and fall of the Roman-Byzantine occupations of the desert are at best misleading. Some of the complex relationships between these populations and their environment are explored below.

ROMAN AGRICULTURAL EXPLOITATION OF NORTH AFRICA

This chapter deals primarily with the Roman and Byzantine desert frontiers. Since most of the Roman agricultural expansion into semiarid regions took place across North Africa, the discussion for the Roman period focuses in part on this territory. Some of the factors to consider are the incentives for investing in farming communities in marginal areas, processes working from the *top down* versus those from the *bottom up*, and the effect of climatic stability versus fluctuation on the success or failure of marginal farming endeavors. Incentives and settlement strategies in desert areas were quite different for the Roman as opposed to those of the Byzantine period in this region.

In examining the role of climate and climate change in the economies of early empires it is important to examine the reasons for the extension of farming zones into marginal areas. These are the regions in which both the risks as well as the expense in terms of resources, time, and energy are high. It follows, then, that the perceived gains should also be equally high. The incentives for farming new territory in agriculturally marginal land can be divided into those from the perspective of the empire core and those as seen from the periphery. Some of the motivation originating at the empire core when the Romans were in control of the desert regions of the southern and eastern Mediterranean regions included an increase in the tax base for the center of imperial control (J.-P. Levy 1967; Sinopoli 1994; Woolf 1992), support for military outposts on the frontier (Garnsey 1978), maintenance for mining operations in the desert (Barker 2002), and the control of preexisting trade routes (Garnsey 1978).

Some of the Roman interests and strategies for farming communities in desert areas of North Africa are well known. After the initial subjugation of a local population, the Romans had a two-fold policy toward the native societies. The peasantry was for the most part left to its own economic pursuits and belief systems, as long as it passively accepted Roman political domination. However, the local ruling elite were the focus of intensive "Romanization" in the material, spiritual, and economic realms (Garnsey 1978). Along with a general strategy of Romanization of the elite segments of societies in conquered regions, the Romans carried with them an economic template that was strongly structured by the traditional Mediterranean subsistence economy (Fulford 1992; J.-P. Levy 1967). This economic ideal was promoted by producers from Italy who were interested in stimulating the demand for Roman-type goods with the ever broadening Romanization of the Mediterranean world, even into non-Mediterranean ecological realms (Woolf 1992).

The geographical expansion of Roman economies led to a seamless adoption of farm-management strategies in Mediterranean vegetation zones under their control and even raised local productivity to the extent that fostered the support of larger populations (Greene 2000:51). This improved agricultural technology allowed Roman society and production systems to expand into the desertic zones. Rather than adapt to local systems of desert food production in the form of animal pastoralism, or look toward the south in the use of native African crops or more xeric crops such as millets, they were able to use the resources of a powerful political infrastructure to extend the ideal Mediterranean farming system far away from its more traditional source (J.-P. Levy 1967).

This was true for landowners of Italian origin as well as the elite landowners stemming from native populations. In North Africa, after a region was conquered, land was given over to Roman colonists, retired army veterans, and the Romanized elite rulers of the native inhabitants. Members of the ruling classes based in Rome were also granted estates in North Africa, and in the later periods of the empire, large tracts of land were appropriated by the emperor (especially Nero) for his own personal profit and gain (Garnsey 1978; Garnsey and Saller 1987:66) Since land in the southern provinces was given to members of the Roman elite society, and as a reward to retired Roman and Romanized military veterans, there was an expansion of the Mediterranean farming economy into the circum-Mediterranean desert zones (Garnsey 1978). Land was taken away from local owners and incorporated into the economy of the Roman sphere of influence. A significant proportion of the produce from this land went back to Rome in the form of taxes. This served the multiple purpose of providing additional revenue for the empire, pacification and control of local populations and their territory, and expansion of the *ideal* of Roman Mediterranean subsistence economies and markets to new regions (Fulford 1992).

From the perspective of the peripheries, one concern in understanding Roman period incentives for the expansion of Mediterranean farming systems into desert regions is whether the marginal farming settlements were the result of *top-down* demands such as the payment of tribute and labor requirements, or *bottom-up* processes driven by local incentive to incorporation into greater political, economic, and prestige networks (Sinopoli 1994). The stimulating effect on local economies of new markets brought by contact with the empire can be a strong incentive to engage in labor-intensive farming in climatically marginal zones when better-watered land is scarce or unavailable (J.-P. Levy 1967; Woolf 1992; Barker 2002). These new markets included local as well as foreign exchange.

Much of the demand for Mediterranean products such as olive oil and wine came from newly urbanized local populations (Woolf 1992). The incentives for trade were grounded in an uneven distribution of these goods, the regular occurrence of both food shortages and overabundances (Garnsey 1988), the increasingly higher proportions of landless citizens and subjects, and the reported high quality of products from certain regions, which lent more prestige to those particular foodstuffs (Rathbone 1983; Woolf 1992). Although there is evidence for interregional trade, long-distance exchange across the empire was not as important as intraregional trade. However, even the local trade invigorated market activity and provided a strong stimulus for cash-crop production on farms. It was one major incentive for expansion into more marginal areas of the southern and eastern edges of the empire.

Another critical outcome of empire-wide trade and exchange was its role in buffering the subsistence systems, and reduction of farming risks for both the Mediterranean zones and the marginal areas around the desert regions. The pan-Mediterranean ties under the control of the Roman Empire greatly expanded the insurance net for local farmers at any point in the basin. Previous to the Roman unification of the Mediterranean basin, populations were often subject to serious food shortages brought on by both environmental disasters and human mismanagement. These food crises commonly led to famine, disease, and social disorders (Garnsey 1988:16). With the extension of an economic network and political control across the Mediterranean basin, there was much more opportunity for goods to flow from regions with surplus to regions with food shortages. This was also made possible by the ease with which goods could be transported by sea routes to ports around the Mediterranean (Garnsey 1988; Fulford 1992).

In addition to the abilities of the Romans to transfer grain from productive to nonproductive regions, there were also provisions for the institution of grain commissioners who could control the storage of grain surplus during years of plenty and redistribute it during years of shortages. The imperial government also had the power to divest wealthy landowners of their stores of grain when they were found to be hoarding supplies during periods of food shortage. These factors, combined with improved agricultural technologies, increased the incentives for expansion in areas that traditionally were less productive for farming.

To summarize, the Roman period expansion of farming into the marginal farming zones of the semiarid regions must be viewed in terms of incentives versus the evaluation of risks. Some of the incentives included the profits of new markets and the granting of this land to members of the population such as veteran soldiers and Italian colonists who were previously landless. Even without reference to the role of climatic change, it is possible to assert that the risks of farming in the marginal environments of the southern and eastern empire were probably lower than at any time previously due primarily to the substantial safety net provided by Roman control of the Mediterranean basin area. This is well illustrated by the expansion of Roman cash-crop farming into the deserts of Libya (Barker; 2002).

BYZANTINE EXPLOITATION OF EASTERN MEDITERRANEAN MARGINAL ZONES

The Romanization of the Mediterranean from the third century BC to the fourth century AD provides an informative example of how farming systems in marginal desert regions are integrated into an economic superstructure that contributes clear incentives for expansion into these areas and reduces the risks of climatic fluctuations, making investments in labor and materials more attractive. The situation in the Late Roman-Byzantine Empire was quite different. By the fourth century AD, the Roman empire had broken down into a deeply fragmented Western Empire and a stronger, more economically viable Eastern Empire whose capital was established in Constantinople (modern-day Istanbul) in 330 AD (Randsborg 1991; Cameron 1993). The spiritual consciousness of the empire had shifted from Roman paganism tempered with local religious cults to the monotheism of Christianity. This new philosophy became a major driving force behind the politics of the empire's power structure and the economy of the entire region.

After 200 AD Italy and much of the Western Empire suffered what some scholars believe was an agricultural recession. There was a reduction of markets with a corresponding weakening of autonomy for towns and their agricultural hinterlands. The western Mediterranean market economy shrunk in both the amount of trade as well as its geographical range. Some writers believe that the relative reduction in Italian-produced commodities was a reflection of rapidly increasing demands for agricultural products in Italian cities that exceeded the ability of Italian producers to supply (Whittaker 1988). The situation for both Africa and the Levant was just the opposite. In the Eastern Empire in general and the Levant in particular there was a great expansion of settlement and agricultural development in the hinterlands. Trade in agricultural products was restricted to the eastern Mediterranean (Randsborg 1991:15).

From 400 to 600 AD there were wealthy settlements throughout the Levant from northern Syria south to the Negev Desert in Israel. A large number of farming villages were inhabited by free peasants, but there were also increasingly greater numbers of large farming enterprises associated with monasteries and villa farmsteads that, as in earlier periods in the western and southern Mediterranean, were owned by elite landowners who used vast numbers of slaves or free tenants (Randsborg 1991; Cameron 1993). These farmsteads were often closed economic systems that produced most of the goods they required (J.-P. Levy 1967:85–86). Major construction works of monumental structures were also typical in many localities in the Eastern Empire in this period. Supporting such enterprises would certainly have required a great quantity of resources (Randsborg 1991:47–48).

The economics of desert settlement in late antiquity is not as well documented as in the Classical Roman period. For the deserts of the Eastern Empire, including the Syrian Desert and the Negev, several factors contributing to the overall prosperity of the region might have led to the greater investment in desert farming. One of these is the relative peace and stability of the region during the fifth century AD (Cameron 1993:182). This stability greatly contrasted with the situation in the Western Empire, which was unsettled both politically and economically. As a result the Eastern Empire began to draw wealthy refugees from the western provinces. These families brought with them significant funds that acted as a boost to the economy of the region and not only contributed to the prosperity of cities of the Levant but also created more of a demand for olive oil, wine, and luxury products.

One of the most important reasons for the economic boom that infused Palestine with capital, major public works, and a greatly increased population was the phenomenon of the Christianization of the Roman Empire. This historical event catapulted Roman Palestine from an obscure and hardly noticeable province of the vast Roman Empire into its new status as the Holy Land. The new position as the spiritual center of the Empire attracted a flow of capital from imperial and other sources with a period of public investment under Constantine and his successors, private investment up to the death of the Empress Eudocia in 460, and a resumption of public investment under Justinian in the sixth century (Avi-Yonah 1958:41; Hirschfeld 1997; Parker 1999; Shereshevski 1991). These were massive infusions of capital for the building of churches and monasteries and the development of religious sites, which contributed to the support of workers, craftspeople, artisans, and other members of the population who would have required a sizeable subsistence base stimulating the agricultural economy of the region (Avi-Yonah 1958; Cameron 1993).

This can be seen archaeologically by a huge jump in numbers of villages and farming communities throughout the Levant (see the discussion and citations in Hirschfeld 1997). These agricultural endeavors extended into the marginal farming regions of the deserts as well. Given this economic stimulation and the increased demand for subsistence goods as well as agricultural luxuries such as fine oil and wines, there was a clear incentive to expand farming into the unexploited territory of the Negev Desert (Rubin 1991; 1996), although this might have been just one among several reasons for the spread of towns to the Negev.

There is much literature on the advanced runoff water technologies that allowed this intensive agricultural production to take place in these areas of very low rainfall (Bruins 1986; Evenari, Shanan, and Tadmor 1982; Lavee 1997), but one of the great debates in the study of climate change and population dynamics of the region is the role of climate in possibly facilitating or even providing the overall incentive for incursions into what is today a forbidding, high-risk zone for sustained agriculture. Before we discuss the role of climate change in the context of Byzantine agricultural pursuits in the Negev, it is necessary to understand the historical background of the Roman-Byzantine settlement in the region.

IMPERIAL EXPLOITATION OF THE NEGEV DESERT

The Negev Desert is located in southern Israel. The landscape generally consists of silty loess hills in the north and rocky hillslopes and entrenched valleys in the central and southern Negev. Today the climate is characterized by hot summers with August temperatures averaging 33°C in Beer Sheva to 40°C in Eilat, and mild winters with January temperatures of 10°C (Beer Sheva) to 16°C (Eilat). As with the rest of the Southern Levant, the summers are dry, and all the precipitation falls in the months of October to May. Rainfall ranges from about 30 mm in the southern region around Eilat to 200 mm in the Beer Sheva region and 400 mm in the coastal area around Gaza (Orni and Efrat 1980). The northern and coastal regions are suitable for dry farming of wheat and barley; however, the rainfall patterns are erratic with much of the yearly average falling in torrential rainfall events. This results in large quantities of runoff, rather than percolation of rainwater into the soils, thus not replenishing groundwater reserves. Under the current conditions of high temperatures, high evapo-transpiration rates, low overall rainfall, and erratic distribution, agriculture in the Negev is a risky enterprise for those intent upon subsistence and cash-farming pursuits. This begs the question of who were the hardy souls who first attempted intensive farming in the Negev, and what motivated them to undertake this risky enterprise.

Historical Background

The first classical era people to settle in the Negev Desert were the Nabateans. The Nabateans began as a confederacy of nomadic tribes inhabiting north Arabia and later moved into Transjordan, Sinai, and the Negev. From the second century BC until 106 AD they succeeded in building a wealthy kingdom, with its capital at Petra. They were able to amass great wealth and power in the region based on their control of the lucrative trade in medicinal herbs, cosmetics, incense, spices, and perfume along routes from the Red Sea to the Mediterranean (Hammond 1973). Settlement in the Negev Desert began as the Nabateans spread northwest from the Northern Hijaz area of the Arabian Peninsula. Their sites consisted of small settlements and way stations for travelers along the spice route. Some of their settlement remains were excavated from the earliest levels at the urban sites of Avdat, Mampsis, and Halutza, and although the Nabateans had agricultural systems near their settlements in what is today southern Jordan, in the Negev there appears to be no such evidence that dates to this period (Rubin 1997; Shereshevski 1991).

In 106 AD the Roman Empire annexed the territory of the Nabateans in an attempt to control the rich trade networks. Unlike the North African peripheries where Roman agriculture was highly developed and represented the *Mediterraneanization* of marginal farming zones, Classical Roman interests in the Negev were almost nonexistent. Some authors suggest that Roman incursions were based primarily on military control of the "Saracen" populations of nomads (Parker 1987; Mayerson 1989), in an effort to secure the eastern frontier. However, Isaac claims there is little evidence for hostilities between Romans and Saracens on the eastern frontier, and the few Roman "forts," such as the one at Avdat, were more likely to have been "road stations" or short-term "police posts" (Isaac 1992:132). The important observation is that the Classical Roman Empire in Palestine was more interested in Romanizing populations who were already settled agriculturalists, and exploiting traditional farming zones, rather than turning new marginal land into agricultural zones for the production of Mediterranean products (Gihon 1980; Mayerson 1989).

The main flourishing of settlement in the Negev took place in what scholars refer to as the "Late Roman" or the "Roman-Byzantine" period. In Israel, it is referred to exclusively as the "Byzantine period." This episode lasted from the first half of the fourth century AD with Christianity becoming the state religion of the Roman Empire after the conversion of Constantine I in 312 AD and the founding of Constantinople in 330 AD, until the mid-seventh century with the Moslem conquests in 643 AD. The changes in settlement and land use had a major impact on the inhabitants of the Negev, which included Nabateans, Hellenized Mediterranean, and Semitic peoples (Negev 1991). There were shifts from nomadism to sedentary settlements, an imperial infrastructure of administration and law was established, large towns were constructed including monumental architecture, and there were major changes in religious thought from local paganism to the Greco-Roman pantheon and finally to Christianity. Along with these profound social changes, sophisticated agricultural techniques were introduced that brought the region into the sphere of the Mediterranean agricultural economy (Rubin 1996; 1997:268).

The introduction of Christianity to the region brought with it great economic prosperity. The sudden increase in population all over Palestine was likewise reflected in the desert regions of the south with an upsurge of population in the Negev towns. The prosperity benefited the Negev by bringing an influx of skilled craftspeople and new investment in the intensification of desert agriculture. In addition to standard subsistence crops such as wheat,

barley, and lentils, the population of the Negev towns was able to produce oil
and wine, and for the first time enter into the economy of Mediterranean agricultural produce (Shereshevski 1991:3). These cities had little social and economic independence. They were intimately tied into the economics, society,
culture, and administration of the empire, under the direct control of the emperor (Shereshevski 1991:5).

Some of the Roman-Byzantine period settlements are still standing in the
Negev today. The visitor who makes a trip to see them will get a glimpse at the
layout and organization of these towns with all the majesty of intact two-story
structures, basilicas (sometimes three churches in one town), mosaic floors,
wine presses, and an extraordinary array of agricultural works in the surrounding landscape (see figures 8.2 and 8.3).

The agricultural installations in the hinterland of the sites are one of the characteristics that make these towns in the desert so impressive. Under natural conditions, after desert rainstorms, large quantities of water run off hillslopes into
the seasonal drainages and out to the sea, leaving very little moisture behind in
the thin soils of the desert landforms. The inhabitants of the Negev towns took
advantage of the phenomenon of desert runoff to acquire large quantities of wa-

FIGURE 8.2
Standing Structures from the Byzantine Site of Shivta in the Negev Desert, Israel
(Photo Credit S. Rosen).

FIGURE 8.3
Agricultural Installations from the Byzantine Site of Shivta in the Negev Desert, Israel
(Photo Credit S. Rosen).

ter for their agricultural fields. In an area that receives less than 100 mm of rainfall per year—an amount less than half the required yearly average for a successful crop of barley—the Roman-Byzantine farmers were able to increase the
effective moisture to almost five times that amount on average (Evenari, Shanan,
and Tadmor 1982; Lavee 1997; Rubin 1997:271).

The techniques they used to increase the agricultural productivity of the
region had their roots in the farming systems of the Mediterranean world but
were adapted to the climate and the landscape of the Negev as early as the Iron
Age. In the pan-Mediterranean area agricultural intensification in hilly areas
took the form of terrace systems that were constructed horizontally along the
slopes, parallel to the valleys. This system worked well to preserve the deep
soils of the slopes and capture water in a series of stepped fields. Rubin (1997)
argues that farmers adapted this typically Mediterranean system to the Negev
by changing the orientation of the terracing to a vertical configuration vis-à-
vis the slope, allowing water to run off from the hillslope more effectively and
to be concentrated on the floodplains of the wadis, held in place by a series of
horizontal check dams across the wadi, at right angles to the flow. This type of
water trap was also a sediment trap, preserving fine-grained material in the

channel and renewing the fields with both fresh silt and moisture. This system was enormously successful. Modern experiments in reconstructed field systems of this nature allowed researchers to produce good harvests not only of cereals, but also of olives and grapes as well (Evenari, Shanan, and Tadmor 1982; Mayerson 1985).

Researchers who study these field systems have demonstrated through field experimentation that this agricultural technology was used to produce much higher yields of a number of different crops than would otherwise have been possible in an area with such low rainfall amounts (Evenari, Shanan, and Tadmor 1982). This allowed the Negev towns to support large populations and enter into the Mediterranean agricultural production economy by producing wine for export as well as internal consumption. Records of yields at the Byzantine period site of Netzana have come down to us from a cache of papyri uncovered at the site. These textual records inform us that wheat, barley, grapes, olives, figs, and dates were cultivated by the local population in the vicinity of the town. Wheat was by far the most important crop and produced yields averaging close to sevenfold that of the initial amount of seed grain planted, or in the neighborhood of 6000 bushels. Mayerson (1967) estimates that this amount could comfortably provide for a population of 1000 people.

However, as mentioned in previous chapters, the use of average yields, like average rainfall, is a poor tool for estimating the productivity of a region and the population that a given landscape can support. Both the Netzana papyri and the experimental farms indicated that there were years in which the rainfall and yields were extremely low, and years in which rainfall was high, resulting in bumper crops. Mayerson (1967) cites Evenari, Shanan, and Tadmor's measurements of Negev rainfall in the years 1961–1965 (see table 8.1). The average rainfall during this period was nearly 100 mm per year, but this number is deceptive. The actual rainfall amounts varied during this time from 57 to 175 mm per year. During the drought years wheat yields were low even with the addition of the runoff water (about 95 kg/du). Alternatively, during the years of heavier rainfall,

Table 8.1.   Rainfall in mm at Byzantine Period Sites of Shivta and Avdat (Evenari, Shanan, and Tadmor 1963, 1964, cited by Mayerson 1967).

Site	1960–1	1961–2	1962–3	1963–4	1964–5	Average
Shivta	102.5	50.6	29.5	143.0	166.0	98.3
Avdat	57.2	51.0	27.7	175.0	161.0	94.2

the fields were extremely productive (about 195 kg/du) (Evenari, Shanan, and Tadmor 1965; Mayerson 1967).

Under the climatic conditions of the present day, there are frequent droughts interspersed with years of abundant rainfall. Given these conditions, one would expect a well-developed system of storage and a program of redistribution on the part of the administrative authorities to help the population survive one, two, or even three consecutive years of drought. As discussed in chapter 6, this was almost certainly the strategy undertaken by the Early Bronze Age polities in the regions further north in the Levant.

Surprisingly, though, excavations at the six Roman-Byzantine urban centers in the Negev revealed no public facilities for grain storage (Shereshevski 1991; P. Fabian, personal communication, 2003). There were only much smaller areas for private household storage in individual dwellings. Storage at the individual family level is not likely to have been enough to supply the household with a subsistence base to overcome two or more consecutive years of drought conditions. This suggests that either the climate during Roman-Byzantine times was significantly wetter than today with rains that were more reliable, or alternatively the Byzantine inhabitants of the Negev were living under a similar rainfall regime as at present but were able to survive as a function of their place within a much larger economic network controlled by the infrastructure of the greater empire as a whole.

EVIDENCE FOR CLIMATE CHANGE IN LATE ANTIQUITY OF THE NEGEV

In the debate over Roman-Byzantine settlement in the desert a number of researchers have claimed there was no climatic change that would have made the land more hospitable for settlement in the Negev (Hirschfeld 1997; Mayerson 1962; Rubin 1989). Alternatively, other researchers maintain that without climate change there could have been little support for the large populations and substantial towns at that time (Bruins 1986; Issar et al. 1992). Those in favor of the claim for more benevolent climatic conditions sometimes use human settlement itself as a proxy for increased rainfall in the region, which of course can lead to a circularity of the argument. Geomorphological studies point to increased alluviation in the streams as an indicator of greater runoff (Cordova 2000; Goldberg 1986; A. Rosen 1986b; Schuldenrein and Clark 2001), although it is possible that during Roman and Byzantine times the increased agricultural activity itself accelerated soil erosion, and the creation of new stream terraces was due solely to this effect.

Other proxies are difficult to read because of the impact of the extensive farming in this heavily populated era. Pollen profiles change at this time, but these are heavily influenced by the replacement of natural forest areas with fields and orchards.

A brief review of some of the climatic data presented in chapter 5 will help put a clearer perspective on the Roman-Byzantine agricultural exploitation of the Negev. The data from Dead Sea levels are often used as a proxy for Negev climate during the Roman-Byzantine period. Although somewhat more informative than the pollen diagrams, they are not without their ambiguities as well. In general a high lake level in the Dead Sea is understood to indicate an overall increase in regional precipitation. This too must be interpreted with a measure of caution. The lake levels are generally indicative of drier and wetter periods, although deforestation from extensive agricultural activities can increase runoff after storms and thus augment the input of water from the catchment into the Dead Sea. A number of researchers report on high lake levels from around 200 BC to 200 AD, roughly corresponding to the Classical Roman period, and again in the fourth century AD at the end of the Roman period, with a drop-off in the early Byzantine period (Bookman et al. 2004; Enzel et al. 2003; Frumkin et al. 1994). It is significant to note that if the dates on these sequences are taken at face value, the reconstruction of high lake levels can be interpreted as inversely correlated with the periods of agricultural expansion in the Negev (S. Rosen 2004).

Historical sources indicate that in the first centuries AD, the crop yields in Palestine were exceptionally high (Feliks 1971; Sperber 1978:60–61). Feliks attributes this productivity to intensive cultivation on a large number of small holdings, although this might also indicate an increase in rainfall. There is a marked agricultural recession recorded around the middle of the third century and a dramatic revival in the fourth century (Sperber 1978:61–64). Some of this is undoubtedly related to changing agricultural economic practices during the Roman occupation of Palestine (Sperber 1978), but it is also probable that there were fluctuations in rainfall during this period as well (Sperber 1978:66–67).

A seemingly more reliable proxy is the isotopic record from sea sediments and speleothems in caves (Bar-Matthews and Ayalon 2004; Schilman et al. 2001). The recent isotope data from the Mediterranean Sea cores is somewhat better dated than other proxies and provides a higher-resolution record of climatic fluctuations. Although these readings give a more precise record of rainfall and temperature changes, they still lack the very fine resolution in dating essential for comparison with archaeological and historical data. Therefore, even these data are inconclusive about the actual effect the fluctuations would have had on intensive farming in the desert during the Roman-Byzantine times.

The $\delta^{18}O$ data from the eastern Mediterranean, combined with isotopic data from the Nahal Soreq Cave (Schilman et al. 2002), seem to contradict the Dead Sea sequence. The difference in the interpretations is due primarily to the problems with correlating the isotope record from Mediterranean Sea cores and cave speleothems on one hand with the Dead Sea sediment sections and the elevations of rafted wood on the other (Frumkin et al. 1994). The sea-core data suggest enriched levels of $\delta^{18}O$ and thus drier conditions from circa 500 BC to 200 AD with short wet phases from circa 200 to 350 AD and circa 450 to 550 AD. The Soreq Cave speleothem data indicate a moist phase from roughly 500 to 900 AD. In spite of these peaks of wet and dry phases during both the Classical Roman and Byzantine periods, it is essential to note that the Soreq Cave data indicate the overall average rainfall at this time was much like that of today (Bar-Matthews and Ayalon 2004). However, it is erroneous to assign a unified climatic phase, whether wet or dry, cool or warm, to the Roman and Byzantine historical periods. The populations living in the Levant at this time witnessed years that were both wet and dry enough to impact agricultural yields, and it was incumbent upon them to deal with these.

Although it is difficult to pinpoint the exact periods of wet and dry years to achieve a satisfactory mesh of historical and climatic data, it is clear from the proxies as well as the historical data that there were a series of wet and dry years throughout the Roman and Byzantine eras. In addition to the interplay of wet and dry years were the economic changes that took place during the Roman and Byzantine periods. Economic shifts in such things as the management of landholdings and prices of commodities in pan-Mediterranean markets (Garnsey 1988; J.-P. Levy 1967:80; Randsborg 1991) could also lead to economic stress that was unrelated to weather conditions or climate. This economic shift resulted in a reduction of productivity that sometimes led to economic stress and could have in some circumstances compounded the effects of years in which rainfall was lower than average or not well distributed.

It is also essential to emphasize a more important aspect of the climatic proxy data than the struggle of trying to match up the boundaries of wet phases with the boundaries of historical periods. Traditionally the debate about the effect of climate change in the Roman-Byzantine period has centered on the possibility that overall rainfall amounts in the past were greater than at the present time. The climatic shifts within these periods highlight the futile nature of a debate based on the simple notion of wet versus dry episodes. There is more to climatic change than the simple increase or decrease in rainfall averages, and other factors such as climatic stability versus rapid fluctuation have a much greater impact on long-term planning and the development of farming systems. In an earlier work, Bar-Matthews, Ayalon, and Kaufman (1998) suggest that the period from circa 1050 BC to 950 AD had the most stable $\delta^{18}O$ and $\delta^{13}C$ values of the past 6500 years, which translate into stable rainfall amounts (see figure 5.3 in chapter 5). This stability surpassed that of today and far exceeded the readings of previous periods in the Holocene oxygen isotope record. It is the stability of rainfall fluctuation that is of critical importance for understanding the farming systems in the desert and far outweighs the importance of absolute quantities of rainfall in marginal farming areas in these periods of sociopolitical complexity.

In times of climatic stability lasting more than a generation, farmers gain the confidence to expand into more risky enterprises that might be too economically tenuous in periods in which there is more noticeable climatic fluctuation and variability on a year-by-year or decade-by-decade basis. A more stable climatic situation encourages farmers to make more of an investment in extending the types of crops grown to include those that can be sold as cash crops. Shorter-term variation in wet versus dry years, the seasonality of the rains, and their impact on farming systems would have a direct economic impact on farming. If drought years came with reliable regularity, then farmers could plan for them and recover more readily than in situations where droughts were less predictable. The seasonality of rains is also an important factor. Even with lower rainfall averages, if rains fall in the appropriate season, this would lend greater stability to the farming systems. The observation by Bar-Matthews, Ayalon, and Kaufman (1998) that the isotopic signal indicating rainfall distribution fluctuated less in the Roman and Byzantine periods than at any time previously or since is of greatest importance for understanding the willingness of the Byzantine society to invest in settlements in the Negev Desert at this time.

INTERPLAY BETWEEN CLIMATE, ENVIRONMENT, AND EMPIRES IN THE NEGEV

It is clear from the above review of climatic data that it is impossible to assign a label of wet or dry and warm or cool to an entire historical period of either Roman or Byzantine settlement in the Southern Levant. Populations living in the region at that time had to contend with a combination of wet and dry episodes that would have impacted their economies in both beneficial and adverse ways. One also cannot ignore the agricultural technology and systems of land allocation in attempting to understand changing environment and its impact on agricultural economies at the time.

In considering the role of climate in the economy of empires it is also important to stress that to a large degree, agricultural expansion into marginal farming zones is far more dependent upon rainfall predictability than it is on changes in yearly rainfall averages. Even given low amounts of rainfall per year, empires are capable of garnering the technological know-how for major feats of agricultural technology, as in the case of Roman aqueduct building in North Africa and Byzantine runoff farming in the Negev Desert. They were capable of transforming sandy wastelands into productive farming enterprises. As long as even low rainfall came in predictable quantities, these ancient empires found it worthwhile to expend energy and resources to farm these desert regions.

To make these efforts worthwhile, there must have been profitable incentives that drew Roman-Byzantine populations into these desert regions. It is clear that they did not arrive in the desert as independent village subsistence farmers in the manner of the occupants of the post-Byzantine Islamic period villages in the Negev (Haiman 1995). There were both "push" and "pull" factors that operated on these communities. The push factor that moved population out of the center of the Palestine and into the desert was in part the large population increase in the Byzantine period and the expansion into the lush agricultural valleys of the more temperate Mediterranean regions of the Southern Levant, making good agricultural land increasingly scarcer. Another push was the ideology of "civilizing" the frontier by bringing the knowledge of Christianity to the nomadic tribal Saracens of the desert lands. These incentives were counterbalanced by a set of pull factors that might have enticed populations into the Negev settlements. These included such aspects as control over trade routes from the Red Sea to the Mediterranean, which crossed the Negev from Aila (Eilat) to Gaza. Another highly significant source of revenue came from overseeing the pilgrim routes from the holy sites in Palestine to the monasteries and places of worship in the Sinai Desert. This inspired what can be understood in modern terms as a lucrative tourist trade and provisioning. Finally, the Byzantine period towns on the frontiers of civilization were also a stabilizing force on the boundaries of the empire.

In addition to these above-mentioned motivations for establishing settlements in the Negev, one of the most important enticements to Negev farmers was the benefits of the region for growing grapes in conditions that would produce exceptionally fine wines. The heat of the growing season concentrated sugars in the grape, and the results were a superior product. Wines from the Negev in general and Gaza in particular were highly prized throughout the empire for their high quality and good taste (Mayerson 1985). To judge from the large numbers of wine presses at Byzantine Negev town sites, wine production was a major cash enterprise for these communities (see figure 8.4).

Taken as a whole, these economic pursuits show how varied the economy of the Byzantine towns in the Negev was. Diversification has always been an essential strategy for populations living in marginal farming zones. In the

FIGURE 8.4
Wine Press and Collection Vat from the Byzantine Town of Shivta in the Negev Desert (Photo Credit S. Rosen).

case of simple subsistence farmers, this manifests itself in the diversification of crop plants and herd animals. For citizens of the Byzantine Empire, with a complex economy tied into world markets, diversification came in the form of a variety of economic pursuits. However, these populations still had to contend with a fluctuating climatic regime in which some years resulted in agricultural failures. As discussed above, these drought years might have been more predictable than in earlier times, or even than today. The people living in the Negev towns had a number of options for survival of these droughts. First, they were able to fall back on the revenue from other sources of income to buy the food and other commodities needed for survival. Second, as a part of the larger empire, they had a significantly large and complex infrastructure that could have provided a buffer of outside support in times of need. The lack of large public storage facilities at these sites suggests that food in times of drought might have come from sources well outside the catchment of these towns, either in the form of food purchased by individual inhabitants of the cities or as relief supplies brought into the region from the heart of the empire.

These incentives were critical for the initiation of Negev settlement by Byzantine populations. However, the human endeavor seems to have gone hand in hand with a climatic regime that fluctuated at a lower amplitude than in previous periods. This sequence of wet and dry years might have posed a barrier to earlier populations that intended to build settlements in the Negev Desert, but for the Byzantine inhabitants it was deemed acceptable to a population with more sophisticated agricultural technology and extensive economic ties across the eastern Mediterranean region and beyond.

This again illustrates the point that there are no simple scenarios vis-à-vis climate and culture. If we want to understand the role of climate in imperial societies, we cannot match up a chart of historical periods on one axis and a list of wet and dry periods on the other, as has been published in many previous works on the subject. The story is a much more complex relationship composed of interactions between a number of different climatic variables (temperature, amount of rainfall, and distribution and fluctuation of rainfall), and a list of cultural variables (technological level, social organization, cosmology, economic ties with world systems, the strength of incentives for marginal farming, and buffering strategies). It is only by outlining these factors that we can approach an understanding of the relationships between empires and the fluctuating climates in which they existed.

Civilizing Climate

This book has attempted to illuminate two different lines of evidence, one from natural science paleoclimatic studies, and the other from social archaeology, anthropology, and history. They have been brought together to outline a narrative of the human experience with climate change through time and to illustrate some key considerations of how human societies respond when suddenly faced with problems of changing climate and environments. The title of the book and this last chapter attempt to emphasize the ambiguity in the reciprocal relationship between human societies and environmental change. On one hand the need to adapt to sometimes rapidly changing climates is one of the challenges that can stimulate technological, social, and political innovation. In this sense climate change can be viewed as a *civilizing factor*. On the other hand, some human groups have become more capable of controlling the effects of climatic change in their particular locality, time frame, and cultural milieu. In some senses these societies can be perceived as a *civilizing force* upon aspects of nature and climate.

The case studies presented here were taken from a region of the eastern Mediterranean that has both good paleoclimatic proxy data and a well-defined archaeological record, although similar conditions are present in many localities around the globe. The three case studies of hunter-gatherers, early complex societies, and empires were deliberately selected to emphasize the very different relationships with the natural world and the responses to climate change among three distinctly different types of social organizations.

There are aspects of these case studies that can be viewed as uniquely individual to their particular environmental and social milieu, but at the same time some elements of their responses to environmental challenges can be used as models for other social groups facing the threat of climate change.

In their elegant introduction to *The Way the Wind Blows*, McIntosh, Tainter, and McIntosh (2000b) point to some decisive factors in human-climate relations. We must not forget that the concept of *adaptation* relates to both technological adjustments as well as changes in human social relationships. Also, human perceptions of climatic change are specific to individual societies and very much influence how a particular population will respond in the face of environmental shifts. This is intimately tied to how a society defines the place of humanity in nature, how climatic change is perceived in terms of its magnitude and duration, and the influence of social memory on planning strategies. Different segments of society have different motivations for action when faced with environmental change. These can result in a diverse range of strategies. Social and technological resilience versus resistance to change can be critical factors in the ability of societies to adapt to new environmental conditions. Resilience may vary between segments of society leading to greater resistance to change among those members of society who benefit from current conditions.

Some of these principles were considered here in the process of constructing models for social responses. In both the paleoclimatic reconstructions and the models of human responses I have presented scenarios of how the relationships might have played out. These are proposed here as likely relationships rather than a prescribed series of events. The main purpose is to stimulate critical thinking on the subject of humans and their changing environments, as a rebuttal to more simplistic models of Pavlovian stimulus/response that appear in some of the paleoclimatic and archaeological literature. In pursuit of this goal, chapter 2 outlines some of the techniques used in reconstructing paleoclimates and how a nonspecialist might evaluate the results and interpretations of these data. These interpretations are not always straightforward, and a measure of critical consideration is required before one accepts a climatic scenario based on any one proxy. I apply these principles in an analytical review of proxy paleoclimatic data for the Pleistocene (chapter 4) and Holocene (chapter 5).

Chapter 4 focuses on the dramatic changes of climate in the eastern Mediterranean region as the world progressed from the height of the Late

Glacial Maximum circa 23,000 BP through the end of the Pleistocene. As discussed in this chapter the progression from very cold and dry conditions toward the current warm Holocene was neither steady nor unidirectional. The
human story in this time period was equally diverse. The Kebaran complex
hunter-gatherer populations that lived in the region at this time were well
adapted to these cool conditions and appear to have had a broad-based economy including the use of steppic resources such as small game, cereals, and
other large-seeded grasses. Other major elements of the climatic history of
this time period include a postglacial warming with moister conditions beginning around 16,000 cal. BP, which allowed the growth and expansion of
Mediterranean woodlands. The Early Natufian populations of northern Israel
were complex hunter-gatherers who grabbed the attention of archaeologists
interested in the origins of agriculture by being the first populations to live in
large sedentary and semisedentary hamlets and small villages. They also had
what appear to be technologies preadapted to an agricultural lifestyle, such as
abundant quantities of flint sickle segments and grinding stones for processing nuts and seeds. They were clearly intensive exploiters of the progenitors of
domestic wheat and barley, as well as other plant resources, presumably including acorns and other nuts.

Natufian society was impacted by the climatic events of the Younger Dryas,
which brought a reversal of the climatic conditions around 13,000 cal. BP with
the return of cool dry conditions. One of the human survival responses to this
environmental shift was the cessation of the trend toward sedentism and increased mobility among Natufian populations of the Late and Final Natufian
periods. Chapter 6 contends that the environmental changes decreased the
plant resources available from trees but increased the stands of large-seeded
grasses, encouraging some Natufian groups to exploit more of this resource in
the same way that other hunter-gatherers from ancient Australia, California,
and the American Great Plains adopted such strategies under similar circumstances. Natufian period populations from other localities in the Levant
adopted other strategies for survival. However, chapter 6 argues that rather
than leading to an agricultural way of life, the strategy of seed exploitation was
so successful that it maintained a Natufian status quo that lasted for over another thousand years.

The Younger Dryas ended with the beginning of the Holocene, which was
marked by a rapidly warming and moister climate. At this time, small villages
began reappearing in the Southern Levant, and these heralded the formal be-

ginning of the Neolithic with the Pre-Pottery Neolithic A period. In spite of its general acceptance as the first Neolithic, PPNA sites have yielded precious little if any hard evidence for cereal cultivation. If we examine the period from a perspective of the *longue durée*, it might be possible to suggest that PPNA societies were back on the track toward sedentism and social complexity begun by the Early Natufians. In the case of the Natufians this trend was interrupted by the onset of the Younger Dryas and its demands for new subsistence adaptations, whereas the PPNA peoples had the benefits of a much more favorable environment within which to develop. At this time human settlements became denser, new ideas and concepts developed, and new subsistence economies eventually led to fully agricultural societies in the PPNB. This agricultural economy was for the most part successful due to the very moist climatic conditions of the Early Holocene, and the PPNB way of life spread rapidly throughout the Near East.

Although this scenario might sound somewhat programmatic, it is suggested as one possible sequence of events. It is very much the philosophy of this volume that when the climate changes dramatically, it is difficult to predict how a particular society will respond. Some responses are successful and appear to be adaptive; others are less so and appear in the archaeological record as abandonments or social collapse. The choices societies make relate to their worldview, their type of social and political structures, as well as their level of technological development. In the above case of the Natufian and Pre-Pottery Neolithic societies, the choices appear to us to follow a straight line of development toward agriculture only because we now know the outcome of that particular historical trajectory. However, there were many hunter-gatherer and early village societies worldwide that were faced with similar push and pull factors, although they had other solutions and other adaptations, and in the end they did not develop large village societies and agriculture. In most cases we may never know the motivations of such populations and why they succeeded or failed. Some of the evidence appears in the tangible settlement and subsistence remains left behind at archaeological sites, while other parts of the record are lost in the ether of human perceptions and beliefs. However, as we approach our own time period, we have more of a chance of glimpsing bits of the story from written records or tangible art objects that allow us a small peek into the social psyche of a particular population.

Chapter 5 examines the evidence for Holocene climatic change. Although for the most part the changes were on a smaller scale than those occurring between glacial and interglacial periods, and therefore not as dramatic, they were certainly intense enough to have impacted human societies, especially when considering that populations at that time were using new social and technological adaptations to push the limits of their carrying capacities, feeding larger populations, and exploiting more marginal environmental zones. The Early Holocene began with significantly warmer and moister conditions as the region pulled out of the preceding cold dry Younger Dryas event. There was a brief reversal coinciding with the Holocene 8.2KY event, and then a return to a moister warmer environment. The Middle Holocene was on average moister than the present-day climate of the Near East; however, it was subject to rapid and profound fluctuations between very wet and very dry climatic conditions. This presented the possibilities of both tremendous development as well as risk of devastation to the nascent towns and early complex societies of the region during the Chalcolithic and Early Bronze Ages. Chapter 7 examines how EB societies responded to this environmental dialectic. This treatment of the topic proposes that in one sense the elite managers of the society were able to take advantage of the feast-or-famine environmental regime by gaining control over the belief systems as well as the economic output of farmer-producers.

The EB II–III towns in the Southern Levant were well adapted to a recurring cycle of droughts through a combination of their agricultural technology and social organization. The agricultural technology took advantage of the environmental and hydrological regime of the period. Naturally aggrading floodplains allowed a farming buffer against drought through floodwater irrigation and exploitation of floodplains for intensive farming of cereals. The social aspects, including taxation and redistribution of goods, were driven by a belief system that placed the welfare of farming in the hands of the elite and priestly classes. The model suggests that cosmology dictated that the elite managers were a direct link to the deities of rain and were responsible for ensuring the fertility of crops. In return, the farmers were indebted to the central authorities and therefore required to pay a tax of grain, which then could be redistributed to the population during times of drought and greatest need. The model suggests that this system functioned with variations on the theme over the course of a millennium. However, weaknesses in the structure were ultimately responsible for its eventual breakdown.

A primary shortcoming was the removal of control over surplus goods from the hands of the subsistence farmers upon which the economic system was based. The farming segment of society was locked into the system by handing over surplus to the storerooms of the elite managers. Therefore, in times of drought they were dependent upon the managers for their survival. There are numerous ethnographic analogies for these circumstances in both the present and past. Drawing further on analogy, the model suggests that elite managers were slow to respond to an environmental crisis at the end of the third millennium BC. This environmental crisis took the form of a changed hydrological regime in which the streams began to incise their beds, leaving the floodplains high and permanently dry. The shift removed the possibility of using floodwater farming as a buffer against drought years. At the same time, the region was hit by severe droughts that heralded the beginning of the drier Late Holocene. As a result the agricultural buffer disappeared at exactly the time when it was most needed.

There is no evidence that populations living in the Southern Levant at the time made attempts to activate public works projects such as improved irrigation systems to alleviate the environmental stress. However, there were signs of increased cultic activity through more temple construction. This is emphasized as a possible example of an inappropriate and unsuccessful response. A general collapse of virtually all EB towns in the region highlights the failure of the society to readjust their economies. The only towns to survive and even thrive were ones located near perennial streams on the east side of the Jordan Valley. The important epilogue to this tale of nonadaptive response is the fact that subsequent populations were able to thrive in the Southern Levant and support large towns and cities under the same very dry conditions and the same hydrological regime of downcutting that seems to have initiated the demise of most of the Early Bronze Age populations of the region. This model proposed in chapter 7 underscores the less-tangible elements of environmental perception that can greatly influence how societies respond to periods of environmental stress. It also emphasizes the sometimes conflicting motivations and interests of different social segments within one society.

The drier climate of the Late Holocene appears to go hand in hand with relatively greater stability throughout much of this phase. The Late Holocene, although more stable than previous periods, does have enough variability to greatly impact human societies as seen in the influence of the Medieval Warm Period versus the Little Ice Age on European civilization in the last 1500 years. Chapter 8 deals with probably the most stable of climatic episodes in the Southern Levant, that of the period from circa 200 BC to circa 600 AD during the period of the Roman and Byzantine imperial expansion into the deserts of North Africa and the Southern Levant. There has always been much controversy over the possibility that the Roman-Byzantine Empire took advantage of moister climatic conditions to agriculturally exploit these traditionally marginal territories. The opposing view is that the environment was similar to that of today, and it was the imperial infrastructure that allowed the exploitation of the deserts at that time.

In a sense, there is merit in both arguments. Although there is compelling climatic evidence from the Dead Sea for moister phases over the course of this time period, there is equally strong evidence from oxygen isotope studies of sea cores and cave speleothems that the overall climatic regime was as dry as or even drier than that of today. The argument in chapter 8 proposes that climate did indeed play an important role in the exploitation of the desert fringes of the Mediterranean during the Roman-Byzantine period. However, this role was not one of sustained high rainfall phases that coincided with these periods. Rather, the element of relative climatic stability was the critical factor that allowed the empires to profit from agricultural incursions into the semiarid zones. Climatic stability allows a society to strategically plan for semiarid agriculture through improvements in dry-land farming technology and preparation for predictable droughts by activating a social safety net with its roots in an economy that could draw upon resources throughout the Mediterranean world and beyond.

The Roman-Byzantine period expansion into the deserts must be viewed in terms of *push-pull* motivations. These are detailed in chapter 8, but in general, some aspects of push factors suggest that the Roman expansion into the semiarid zones of North Africa are in some senses a result of the spread of a Mediterranean economic and conceptual template among foreign populations at the fringes of the empire. The Mediterranean economy was one basic element defining perceptions of the civilized versus the barbaric world. Other aspects include economic incentives of expanding markets of the empire and the provisioning of the capital of Rome itself. Climatic stability was one component of the pull factor.

The motivations for the Byzantines to invest in major towns and extensive agricultural systems in the Negev Desert were quite different. Some were intimately tied with early Christian ideology and the relatively sudden focus of attention on the Southern Levant as a center of spiritual inspiration. This attracted increasingly large populations into the region and led to an eco-

nomic boom based on the pilgrim trade as well as on provisioning the growing population. Importantly, the relative political stability of the Eastern Empire attracted wealthy refugees from the more turbulent Western Empire,
which also added to the growth of the economy. The strength of the economy
reduced the risk of farming marginal zone lands and allowed for a buffer that
was supported by the infrastructure of the Byzantine Empire. The investment
in extensive and sophisticated agricultural systems was worth the payoff in the
form of stability of the desert fringes, increased wine and olive oil production,
as well as the maintenance of the pilgrim trail. Again, the more predictable climate was an asset to these economic and social endeavors.

This volume has used examples of ancient Near Eastern societies at different levels of social, political, and technological organization to illustrate some
key aspects of human-environmental relations. These past societies had very
different social organizations, economies, belief systems, and ways of perceiv-
ing the world that certainly influenced how they prepared for and reacted to
times of environmental crisis and opportunity. They are not direct analogies
to other societies worldwide, but perhaps they serve as an example of ways to
examine and study the reactions of past societies to environmental change.

Although this book has dealt with human communities from the distant
and perhaps murky past, there are certainly lessons that can be informative for
us today, as modern societies are attempting to cope with the realities of
changing climate and our ever-shifting relationships with the natural world.
Today, as in the past, much of the story is about the way people perceive their
relationship with nature. This is a difficult topic to analyze for past societies in
spite of its pivotal role in understanding human-nature relations, societies'
shaping of the natural world, and ultimately their survival. But the importance of perceptions of nature and climate change should not be lost on scholars of the modern world who attempt to understand how our own Western
society responds to and prepares for climatic change and natural disasters of
great magnitude. There is a sense that the modern Western world is couched
within an overly self-assured attitude toward nature and climate change. For
people of faith, this attitude dictates that God and by extension nature is on
"our" side. For people of science and technology, there is an unfailing belief
that our instruments and gadgets are ever prepared to help us overcome all
environmental trials. This overconfidence makes it all the more startling at
first, and finally poignant, that the responses of some of the richest and most
powerful societies of our modern world can crumble so thoroughly in the face

of natural disasters, as witnessed by the devastation and slow recovery from Hurricane Katrina on the Gulf Coast of the southern United States in September 2005.

The politics of global warming is another arena for the battles between conflicting perceptions. The proponents of reduction in greenhouse gases take a stand against political and economic conservatism and a reluctance to invest in radical economic changes. Each side in this polemic struggles to influence public opinion, laying its own claim to *truth*. If we can take any guidance from the experience of past societies, it might be in learning the advantages of open-mindedness, economic cooperation, diversity, and resilience in the face of environmental change. To take informed action the public must be aware of the scientific debates concerning environmental change and maintain a critical awareness of the motivations behind the rhetoric of our modern-day *priestly classes* in the form of politicians and industrialists. With the help of paleoclimatic studies, anthropology, archaeology, geography, history, and other such disciplines, a more complete understanding of the past can, we hope, aid us in preparing for a future of uncertain climatic and environmental conditions.

References

Adams, E. C. *The Origin and Development of the Pueblo Katsina Cult.* Tucson: University of Arizona Press, 1991.

Alley, R. B., P. A. Mayewski, T. Sowers, M. Stuiver, K. C. Taylor, and P. U. Clark. "Holocene Climatic Instability: A Prominent, Widespread Event 8200 Yr Ago." *Geology* 25, no. 6 (1997): 483–486.

Amiran, D. H. K. Unpublished data, n.d.

Amiran, R. "A Cult Stela from Arad." *Israel Exploration Journal* 22 (1972): 86–88.

Avi-Yonah, M. "The Economics of Byzantine Palestine." *Israel Exploration Journal* 8, no. 1 (1958): 39–51.

Bar-Adon, P. "Rujm El-Bahr." *'Atiqot, Special Issue: Excavations in the Judean Desert* 9 (1989): 3–14 (in Hebrew).

Barker, G. "A Tale of Two Deserts: Contrasting Desertification Histories on Rome's Desert Frontiers." *World Archaeology* 33, no. 3 (2002): 488–507.

Barker, G., and D. Gilbertson, eds. *The Archaeology of Drylands: Living at the Margin.* London: Routledge, 2000.

Bar-Matthews, M., and A. Ayalon. "Speleothems as Palaeoclimate Indicators: A Case Study from Soreq Cave Located in the Eastern Mediterranean Region, Israel." Pp. 363–391 in *Past Climate Variability through Europe and Africa*, edited by R. Battarbee, F. Gasse, and C. E. Stickley. Dordrecht, Neth.: Springer, 2004.

Bar-Matthews, M., A. Ayalon, and A. Kaufman. "Late Quaternary Paleoclimate in the Eastern Mediterranean Region from Stable Isotope Analysis of Speleothems at Soreq Cave, Israel." *Quaternary Research* 47 (1997): 155–168.

———. "Middle to Late Holocene (6,500 Yr. Period) Paleoclimate in the Eastern Mediterranean Region from Stable Isotopic Composition of Speleothems from Soreq Cave, Israel." Pp. 203–214 in *Water, Environment and Society in Times of Climatic Change*, edited by A. Issar, N. Brown, and Z. Shilony. New York: Kluwer Press, 1998.

Bar-Matthews, M., A. Ayalon, A. Kaufman, and G. J. Wasserburg. "The Eastern Mediterranean Paleoclimate as a Reflection of Regional Events: Soreq Cave, Israel." *Earth and Planetary Science Letters* 166, no. 1–2 (1999): 85–95.

Bartov, Y., S. L. Goldstein, M. Stein, and Y. Enzel. "Catastrophic Arid Episodes in the Eastern Mediterranean Linked with the North Atlantic Heinrich Events." *Geology* 31, no. 5 (2003): 439–442.

Bartov, Y., M. Stein, Y. Enzel, A. Agnon, and Z. Reches. "Lake Levels and Sequence Stratigraphy of Lake Lisan, the Late Pleistocene Precursor of the Dead Sea." *Quaternary Research* 57, no. 1 (2002): 9–21.

Baruch, Uri. "The Late Holocene Vegetational History of Lake Kinneret (Sea of Galilee), Israel." *Paléorient* 12, no. 2 (1986): 37–48.

———. "Palynological Evidence of Human Impact on the Vegetation as Recorded in Late Holocene Lake Sediments in Israel." Pp. 283–293 in *Man's Role in Shaping of the Eastern Mediterranean Landscape*, edited by S. Bottema, G. Entjes-Nieborg, and W. van Zeist. Rotterdam: Balkema, 1990.

———. "The Late Quaternary Pollen Record of the Near East." Pp. 103–119 in *Late Quaternary Chronology and Paleoclimates of the Eastern Mediterranean*, edited by O. Bar-Yosef and R. S. Kra. Tucson: *Radiocarbon*, University of Arizona Press, 1994.

Baruch, U., and S. Bottema. "Palynological Evidence for Climatic Changes in the Levant ca. 17,000–9,000 Bp." Pp. 11–20 in *The Natufian Culture in the Levant*, edited by O. Bar-Yosef and F. R. Valla. Ann Arbor, MI: International Monographs in Prehistory, 1991.

———. "A New Pollen Diagram from Lake Hula: Vegetational, Climatic and Anthropogenic Implications." Pp. 75–86 in *Ancient Lakes: Their Cultural and Biological Diversity*, edited by H. Kawanabe, G. W. Coulter and A. C. Roosevelt. Brussels: Kenobi, 1999.

Bar-Yosef, Ofer. "Prehistory of the Jordan Rift." *Israel Journal of Earth Sciences* 36 (1987): 107–119.

———. "The Archaeology of the Natufian Layer at Hayonim Cave." Pp. 81–92 in *The Natufian Culture in the Levant*, edited by O. Bar-Yosef and F. Valla. Ann Arbor, MI: International Monographs in Prehistory, 1991.

———. "The Impact of Late Pleistocene–Early Holocene Climatic Changes on Humans in Southwest Asia." Pp. 61–78 in *Humans at the End of the Ice Age: The Archaeology of the Pleistocene-Holocene Transition*, edited by L. G. Straus, B. V. Eriksen, J. M. Erlandson, and D. R. Yesner. New York: Plenum, 1996.

———. "Prehistoric Chronological Framework." Pp. xiv–xvi in *The Archaeology of Society in the Holy Land*, edited by T. E. Levy. London: Leicester University Press, 1998.

———. "The World around Cyprus: From Epi-Paleolithic Foragers to the Collapse of the PPNB Civilization." Pp. 129–164 in *The Earliest Prehistory of Cyprus: From Colonization to Exploitation*, edited by S. Swiny. Boston: American Schools of Oriental Research, 2001.

———. "Natufian: A Complex Society of Foragers." Pp. 91–149 in *Beyond Foraging and Collecting: Evolutionary Change in Hunter-Gatherer Settlement Systems*, edited by B. Fitzhugh and J. Habu. New York: Kluwer, 2002a.

———. "The Natufian Culture and the Early Neolithic: Social and Economic Trends in Southwestern Asia." Pp. 113–126 in *Examining the Farming/Language Dispersal Hypothesis*, edited by P. Belwood and C. Renfrew. Cambridge: University of Cambridge, 2002b.

Bar-Yosef, O., and A. Belfer-Cohen. "The Origins of Sedentism and Farming Communities in the Levant." *Journal of World Prehistory* 3 (1989): 447–498.

———. "From Sedentary Hunter-Gatherers to Territorial Farmers in the Levant." Pp. 181–202 in *Between Bands and States*, edited by S. A. Gregg. Carbondale, IL: Center for Archaeological Investigations, 1991.

———. "From Foraging to Farming in the Mediterranean Levant." Pp. 21–48 in *Transitions to Agriculture in Prehistory*, edited by A. B. Gebauer and T. D. Price. Madison, WI: Prehistory Press, 1992.

———. "Facing Environmental Crisis: Societal and Cultural Changes at the Transition from the Younger Dryas to the Holocene in the Levant." Pp. 55–66 in *The Dawn of Farming in the Near East*, edited by R. T. J. Cappers and S. Bottema. Berlin: Ex Oriente, 2002.

Bar-Yosef, O., P. Goldberg, and T. Leveson. "Late Quaternary Stratigraphy and Prehistory of Wadi Fazael, Jordan Valley: A Preliminary Report." *Paléorient* 2 (1974): 415–428.

Bar-Yosef, O., and M. E. Kislev. "Early Farming Communities in the Jordan Valley." Pp. 632–642 in *Foraging and Farming: The Evolution of Plant Exploitation*, edited by D. R. Harris and G. C. Hillman. London: Unwin Hyman, 1989.

Bar-Yosef, O., and R. S. Kra. *Late Quaternary Chronology and Paleoclimates of the Eastern Mediterranean*. Tucson: University of Arizona Press, 1994.

Bar-Yosef, O., and F. R. Valla, eds. *The Natufian Culture in the Levant*. Ann Arbor, MI: International Monographs in Prehistory, 1991.

Baumgarten, Y. *Archaeological Survey of Israel Map of Shivta (166)*. Jerusalem: Israel Antiquities Authority, 2004.

Begin, Z. B., W. Broecker, B. Buchbinder, Y. Druckman, A. Kaufman, M. Magaritz, and D. Neev. *Dead Sea and Lake Lisan Levels in the Last 30,000 Years: A Preliminary Report*. Jerusalem: Geological Survey of Israel, 1985.

Begin, Z. B., A. Ehrlich, and Y. Nathan. "Lake Lisan: The Pleistocene Precursor of the Dead Sea." *Bulletin of the Geological Survey of Israel* 63 (1974): 0–32.

———. "Stratigraphy and Facies Distribution in the Lisan Formation: New Evidence from the Area South of the Dead Sea, Israel." *Israel Journal of Earth Sciences* 29 (1980): 182–189.

Ben-Tor, A. "The Trade Relations of Palestine in the Early Bronze Age." *Journal of the Economic and Social History of the Orient* 29 (1986): 1–27.

———. *The Archaeology of Ancient Israel*. New Haven, CT: Yale University Press, 1992.

Berlin, A. M. "The Hellenistic Period." Pp. 418–433 in *Near Eastern Archaeology: A Reader*, edited by S. Richard. Winona Lake, IN: Eisenbrauns, 2003.

Bienkowski, P. "Prosperity and Decline in LBA Canaan: A Reply to Leibowitz and Knapp." *Bulletin of the American School of Oriental Research* 275 (1989): 59–61.

Binford, L. R. "Post-Pleistocene Adaptations." Pp. 313–341 in *New Perspectives in Archaeology*, edited by L. R. Binford and S. R. Binford. Chicago: Aldine, 1968.

Bookman, R., Y. Enzel, A. Agnon, and M. Stein. "Late Holocene Lake Levels of the Dead Sea." *Geological Society of America Bulletin* 116, no. 5/6 (2004): 555–571.

Borowski, O. *Agriculture in Iron Age Israel: The Evidence from Archaeology and the Bible*. Winona Lake, IN: Eisenbrauns, 1987.

Bottema, S. "The Use of Palynology in Tracing Early Agriculture." Pp. 27–38 in *The Dawn of Farming in the Near East*, edited by R. T. J. Cappers and S. Bottema. Berlin: Ex Oriente, 2002.

Bottema, S., G. Entjes-Neiborg, and W. van Zeist, eds. *Man's Role in the Shaping of the Eastern Mediterranean Landscape.* Rotterdam: Balkema, 1990.

Bradley, R. S. *Paleoclimatology: Reconstructing Climates of the Quaternary.* New York: Academic Press, 1999.

Braidwood, R. J. B. "From Cave to Village in Prehistoric Iraq." *American Schools of Oriental Research* 124 (1951): 12–18.

———. *Prehistoric Men.* Chicago: Chicago Natural History Museum, 1961.

Brown, N. *The Impact of Climate Change: Some Indications from History, AD 250–1250.* Oxford, UK: Oxford Centre for the Environment, Ethics and Society, 1995.

Bruins, H. J. *Desert Environment and Agriculture in the Central Negev and Kadesh-Barnea during Historical Times.* Nijkerk, Neth.: MIDBAR Foundation, 1986.

Bruins, H. J. The Impact of Man and Climate on the Central Negev and Northeastern Sinai Deserts during the Late Holocene. Pp. 87–99 in *Man's Role in the Shaping of the Eastern Mediterranean Landscape,* edited by S. Bottema, G. Entjes-Neiborg, and W. van Zeist. Rotterdam: Balkema, 1990.

Bunimovitz, S. "On the Edge of Empires: Late Bronze Age (1500–1200 BCE)." Pp. 320–331 in *The Archaeology of Society in the Holy Land,* edited by T. E. Levy. London: Leicester University Press, 1998.

Butzer, K. W. *Environment and Archaeology: An Ecological Approach to Prehistory.* Chicago: Aldine, 1971.

———. "Patterns of Environmental Change in the Near East during Late Pleistocene and Early Holocene Times." Pp. 389–410 in *Problems in Prehistory: North Africa and the Levant,* edited by F. Wendorf and A. E. Marks. Dallas: Southern Methodist University Press, 1975.

———. "Holocene Alluvial Sequences: Problems of Dating and Correlation." Pp. 131–141 in *Time-Scales in Geomorphology,* edited by J. Lewin, D. Davidson, and R. Cullingford. New York: Wiley & Sons, 1980.

———. *Archaeology as Human Ecology: Method and Theory for a Contextual Approach.* Cambridge: Cambridge University Press, 1982.

Byrd, B. F. "Late Quaternary Hunter-Gatherer Complexes in the Levant between 20,000 and 10,000 BP." Pp. 205–226 in *Late Quaternary Chronology and Paleoclimates of the Eastern Mediterranean,* edited by O. Bar-Yosef and R. S. Kra. Tucson: University of Arizona Press, 1994.

Callaway, J. A. *The Early Bronze Age Citadel and Lower City at Ai (Et-Tell): Report of the Joint Archeological Expedition to Ai (Et-Tell), No. 2.* Cambridge, MA: American Schools of Oriental Research, 1980.

Cameron, A. *The Mediterranean World in Late Antiquity, AD 395–600.* London: Routledge, 1993.

Cappers, R. T. J., S. Bottema, and H. Woldring. "Problems in Correlating Pollen Diagrams of the Near East: A Preliminary Report." Pp. 160–169 in *The Origins of Agriculture and Crop Domestication,* edited by A. B. Damania, J. Valkon, G. Willcox, and C. Q. Qualset. Rome: International Plant Genetic Resources Institute, 1998.

Cappers, R. T. J., S. Bottema, H. Woldring, H. van der Plicht, and H. J. Streurman. "Modeling the Emergence of Farming: Implications of the Vegetation Development in the Near East during the Pleistocene-Holocene Transition." Pp. 3–14 in *The Dawn of Farming in the Near East*, edited by R. T. J. Cappers and S. Bottema. Berlin: Ex Oriente, 2002.

Carroll, R. P. *Jeremiah*. London: SCM, 1986.

Childe, V. G. *The Most Ancient East*. London: Routledge and Kegan Paul, 1926.

———. *New Light on the Most Ancient East*, 4th ed. London: Routledge, 1952.

Cohen, M. N. *The Food Crisis in Prehistory: Overpopulation and the Origins of Agriculture*. New Haven, CT: Yale University Press, 1977.

Cohen, R. *Archaeological Survey of Israel, Sde Boqer East, Map 168*. Jerusalem: Israel Department of Antiquities, 1981.

———. *Archaeological Survey of Israel, Sde Boqer West, Map 167*. Jerusalem: Israel Department of Antiquities, 1985.

Colledge, S. "Identifying Pre-Domestication Cultivation Using Multivariate Analysis." Pp. 121–131 in *The Origins of Agriculture and Crop Domestication*, edited by A. B. Damania, J. Valkoun, G. Willcox, and C. O. Qualset. Rome: International Plant Genetic Resources Institute, 1998.

Colledge, S., J. Conolly, and S. Shennan. "Archaeobotanical Evidence for the Spread of Farming in the Eastern Mediterranean." *Current Anthropology* 45, Supplement (2004): S35–S58.

Colson, E. "In Good Years and in Bad: Food Strategies of Self-Reliant Societies." *Journal of Anthropological Research* 35 (1979): 18–29.

Colt, H. Dunscombe, ed. *Excavations at Nessana, Auja Hafir, Palestine, Vol. 1*. London: British School of Archeology, 1962.

Cooperative Holocene Mapping Project. "Climatic Changes of the Last 18,000 Years: Observations and Model Simulations." *Science* 241 (1988): 1043–1052.

Cordova, C. E. "Geomorphological Evidence of Intense Prehistoric Soil Erosion in the Highlands of Central Jordan." *Physical Geography* 21, no. 6 (2000): 538–567.

Courty, M. "Le Cadre Paléogeographique des Occupations Humaines dans le Bassin du Haut-Khabur (Syrie du Nord-Est). Premiers Resultats." *Paleorient* 20 (1994): 21–59.

Cronin, T. M. *Principles of Paleoclimatology*. New York: Columbia University Press, 1999.

Crumley, C. L., ed. *Historical Ecology: Cultural Knowledge and Changing Landscapes*. Santa Fe, NM: School of American Research, 1994.

———, ed. *New Directions in Anthropology and Environment: Intersections*. Walnut Creek, CA: Altamira, 2001.

Dalfes, H. N., G. Kukla, and H. Weiss. *Third Millennium BC Climate Change and Old World Collapse*. Berlin: Springer-Verlag, 1997.

Danin, A. *Desert Vegetation of Israel and Sinai*. Jerusalem: Cana, 1983.

Darmon, F. "Essai de Reconstitution Climatique de L'epipaléolithique au Néolithique Ancien dans la Basse Vallée du Jourdain." *Compte-Rendu de l'Académie des Sciences* 307 (1988): 677–682.

Dean, J. S. "Complexity Theory and Sociocultural Change in the American Southwest." Pp. 89–118 in *The Way the Wind Blows: Climate, History, and Human Action*, edited by R. J. McIntosh, J. A. Tainter, and S. K. McIntosh. New York: Columbia University Press, 2000.

deMenocal, P. B. "Cultural Responses to Climate Change during the Late Holocene." *Science* 292, no. 5517 (2001): 667–673.

Dever, W. G. "Social Structure in Palestine in the Iron II Period on the Eve of Destruction." Pp. 416–431 in *The Archaeology of Society in the Holy Land*, edited by T. E. Levy. London: Leicester University Press, 1998.

———. "Chronology of the Southern Levant." Pp. 82–87 in *Near Eastern Archaeology: A Reader*, edited by S. Richard. Winona Lake, IN: Eisenbrauns, 2003.

Dincauze, D. F. *Environmental Archaeology: Principles and Practice.* Cambridge: Cambridge University Press, 2000.

Dirks, R. "Social Responses during Severe Food Shortages and Famine." *Current Anthropology* 21, no. 1 (1980): 21–44.

Druckman, Y., M. Magaritz, and A. Sneh. "The Shrinking of Lake Lisan, as Reflected by the Diagenesis of Its Marginal Oolitic Deposits." *Israel Journal of Earth Sciences* 36 (1987): 101–106.

Dubowski, Y., J. Erez, and M. Stiller. "Isotopic Paleolimnology of Lake Kinneret." *Limnology and Oceanography* 48, no. 1 (2003): 68–78.

Eastwood, W. J., N. Roberts, H. F. Lamb, and J. C. Tibby. "Holocene Environmental Change in Southwest Turkey: A Palaeoecological Record of Lake and Catchment-Related Changes." *Quaternary Science Reviews* 18, no. 4–5 (1999): 671–695.

Edwards, P. C., S. J. Bourke, S. M. Colledge, J. Head, and P. G. Macumber. "The Late Pleistocene Prehistory in the Wadi Al-Hammeh, Jordan Valley." Pp. 525–565 in *The Prehistory of Jordan: The State of Research in 1986*, edited by A. N. Garrard and H. G. Gebel. Oxford: British Archaeological Reports, International Series, 396, 1988.

Egan, T. "Dry High Plains Are Blowing Away, Again." New York Times Internet Version. 2002. www.nytimes.com/2002/05/03/national/03DUST.html (3 May 2002). Retrieved July 2002.

El-Moslimany, A. P. "The Late Quaternary Vegetational History of the Zagros and Taurus Mountains in the Regions of Lake Mirabad, Lake Zeribar and Lake Van." Pp. 343–350 in *Palaeoclimates, Palaeoenvironments and Human Communities in the Eastern Mediterranean Region in Later Prehistory*, edited by J. L. Bintliff and W. van Zeist. Oxford: British Archaeological Reports, International Series 133, 1982.

Emiliani, C. "Pleistocene Temperatures." *Journal of Geology* 63 (1955): 538–575.

———. "Quaternary Paleotemperatures and the Duration of the High Temperature Intervals." *Science* 178 (1972): 398–401.

Enzel, Y., R. Bookman (Ken Tor), D. Sharon, H. Gvirtzman, U. Dayan, Baruch Ziv, and Mordechai Stein. "Late Holocene Climates of the Near East Deduced from Dead Sea Level Variations and Modern Regional Winter Rainfall." *Quaternary Research* 60, no. 3 (2003): 263–273.

Escobar, A. "After Nature: Steps to an Antiessentialist Political Ecology." *Current Anthropology* 40, no. 1 (1999): 1–30.

Esse, D. *Subsistence, Trade, and Social Change in Early Bronze Age Palestine.* Chicago: University of Chicago Press, 1991.

Evenari, M., L. Shanan, and N. H. Tadmor. *Runoff-Farming in the Negev Desert of Israel: Progress Report on the Avdat and Shivta Farm Projects.* Rehovot, Israel: National and University Institute of Agriculture, 1965.

———. *The Negev: The Challenge of a Desert.* Cambridge, MA: Harvard University Press, 1982.

Faegri, K., and J. Iversen. *Textbook of Pollen Analysis.* New York: John Wiley, 1989.

Fagan, B. M. *The Little Ice Age: How Climate Made History, 1300–1850*. New York: Basic Books, 2000.

Feliks, J. F. "Agriculture." *Encyclopaedia Judaica* 2 (1971): 381–398.

Finet, A. "Mari dans son Contexte Geographiue." *MARI* 4 (1985): 41–44.

Finkelstein, I. "The Great Transformation: The 'Conquest' of the Highlands Frontiers and the Rise of the Territorial States." Pp. 349–367 in *The Archaeology of Society in the Holy Land*, edited by T. E. Levy. London: Leicester University Press, 1998.

Finkelstein, I., and R. Gophna. "Settlement, Demographic, and Economic Patterns in the Highlands of Palestine in the Chalcolithic and Early Bronze Periods and the Beginning of Urbanization." *Bulletin of the American School of Oriental Research* 289 (1993): 1–22.

Flannery, K. V. "Origins and Ecological Effects of Early Domestication in Iran and the Near East." Pp. 73–100 in *The Domestication and Exploitation of Plants and Animals*, edited by P. J. Ucko and G. W. Dimblebey. Chicago: Aldine, 1969.

———. "The Cultural Evolution of Civilizations." *Annual Review of Ecology and Systematics* 3 (1972): 399–426.

Flint, R. F. *Glacial and Quaternary Geology*. New York: Wiley, 1971.

Folsom, F. *Indian Uprising on the Rio Grande: The Pueblo Revolt of 1680*. Albuquerque: University of New Mexico, 1973.

Frumkin, A. "The Holocene History of Dead Sea Levels." Pp. 237–248 in *The Dead Sea: The Lake and Its Setting*, edited by T. M. Niemi, Z. Ben-Avraham, and J. R. Gat. Oxford: Oxford University, 1997.

Frumkin, A., I. Carmi, I. Zak, and M. Magaritz. "Middle Holocene Environmental Change Determined from the Salt Caves of Mount Sedom, Israel." Pp. 315–332 in *Late Quaternary Chronology and Paleoclimates of the Eastern Mediterranean*, edited by O. Bar-Yosef and R. S. Kra. Tucson: University of Arizona Press, 1994.

Frumkin, A., D. C. Ford, and H. P. Schwarcz. "Continental Oxygen Isotopic Record of the Last 170,000 Years in Jerusalem." *Quaternary Research* 51 (1999): 317–327.

Frumkin, A., M. Magarit, I. Carmi, and I. Zak. "The Holocene Climatic Record of the Salt Caves of Mount Sedom, Israel." *The Holocene* 1 (1991): 191–200.

Fulford, M. "Territorial Expansion and the Roman Empire." *World Archaeology* 23, no. 3 (1992): 294–305.

Gallant, T. W. "Crisis and Response: Risk-Buffering Behavior in Hellenistic Greek Communities." *Journal of Interdisciplinary History* 19 (1989): 393–413.

Garcia, R. *Drought and Man: Nature Pleads Not Guilty*. Oxford, UK: Pergamon, 1981.

Garnsey, P. D. A. "Rome's African Empire under the Principate." Pp. 223–254 in *Imperialism in the Ancient World*, edited by P. D. A. Garnsey and C. R. Whittaker. Cambridge: Cambridge University Press, 1978.

———. *Famine and Food Supply in the Graeco-Roman World: Responses to Risk and Crisis*. Cambridge: Cambridge University Press, 1988.

Garnsey, P., and R. Saller. *The Roman Empire: Economy, Society and Culture*. London: Duckworth, 1987.

Garrod, D. A. "A New Mesolithic Industry: The Natufian of Palestine." *Journal of the Royal Anthropological Institute* 62 (1932): 257–269.

———. "The Natufian Culture: The Life and Economy of a Mesolithic People in the Near East." *Proceedings of the British Academy* 43 (1957): 211–277.

Gasse, F. "Hydrological Changes in the African Tropics since the Last Glacial Maximum." *Quaternary Science Reviews* 19, no. 1–5 (2000): 189–211.

Geyer, B. "Géomorphologie et Occupation du Sol de la Moyenne Vallée de l'Euphrate dans la Région de Mari." *MARI* 4 (1985): 27–39.

Geyh, M. A. "The Paleohydrology of the Eastern Mediterranean." Pp. 131–145 in *Late Quaternary Chronology and Paleoclimates of the Eastern Mediterranean*, edited by O. Bar-Yosef and R. S. Kra. Tucson: University of Arizona Press, 1994.

Gihon, M. "Research on the Limes Palaestina: A Stocktaking." Pp. 843–864 in *Roman Frontier Studies 1979*, edited by W. S. Hanson and L. J. F. Keppie. Oxford: British Archaeological Reports, International Series 71, 1980.

Gilead, I. "The Chalcolithic Period in the Levant." *Journal of World Prehistory* 2 (1988): 397–443.

———. "The Neolithic-Chalcolithic Transition and the Qatifian of the Northern Negev and Sinai." *Levant* 22 (1990): 47–63.

Godwin, H. *The History of the British Flora: A Factual Basis for Phytogeography.* Cambridge: Cambridge University Press, 1956.

Goldberg, P. "Late Quaternary Environmental History of the Southern Levant." *Geoarchaeology: An International Journal* 1 (1986): 225–244.

———. "Geology and Stratigraphy of Shiqmim." Pp. 35–43 in *Shiqmim I, Studies Concerning Chalcolithic Societies in the Northern Negev, Desert, Israel (1982–1984)*, edited by T. E. Levy. Oxford: British Archaeological Reports, International Series, 356, 1987.

———. "Interpreting Late Quaternary Continental Sequences in Israel." Pp. 89–102 in *Late Quaternary Chronology and Paleoclimates of the Eastern Mediterranean*, edited by O. Bar-Yosef and R. S. Kra. Tucson: University of Arizona Press, 1994.

Goldberg, P., and O. Bar-Yosef. "Environmental and Archaeological Evidence for Climatic Change in the Southern Levant and Adjacent Areas." Pp. 399–414 in *Paleoclimates, Paleoenvironments and Human Communities in the Eastern Mediterranean Region in Later Prehistory*, edited by J. L. Bintliff and W. van Zeist. Oxford: British Archaeological Reports, 1982.

———. "The Effect of Man on Geomorphological Processes Based upon Evidence from the Levant and Adjacent Areas." Pp. 71–85 in *Man's Role in the Shaping of the Eastern Mediterranean Landscape*, edited by S. Bottema, G. Entjes-Neiborg, and W. Van Zeist. Rotterdam: Balkema, 1990.

Goldberg, P., and A. M. Rosen. "Early Holocene Palaeoenvironments of Israel." Pp. 22–33 in *Shiqmim I, Studies Concerning Chalcolithic Societies in the Northern Negev Desert, Israel (1982–1984)*, edited by T. E. Levy. Oxford: British Archaeological Reports, 1987.

Goodfriend, G. A. "Chronostratigraphic Studies of Sediments in the Negev Desert, Using Amino Acid Epimerization Analysis of Land Snail Shells." *Quaternary Research* 28 (1987): 374–392.

———. "Mid-Holocene Rainfall in the Negev Desert from $\delta^{13}C$ of Land Snail Shell Organic Matter." *Nature* 333 (1988): 757–760.

———. "Holocene Trends in $\delta^{18}O$ in Land Snail Shells from the Negev Desert and Their Implications for Changes in Rainfall Source Areas." *Quaternary Research* 35 (1991): 417–426.

————. "Terrestrial Stable Isotope Records of Late Quaternary Paleoclimates in the Eastern Mediterranean Region." *Quaternary Science Reviews* 18, no. 4–5 (1999): 501–513.

Goodfriend, G. A., and Mordeckai Magaritz. "Carbon and Oxygen Isotope Composition of Shell Carbonate of Desert Land Snails." *Earth and Planetary Science Letters* 86 (1987): 377–388.

————. "Paleosols and Late Pleistocene Rainfall Fluctuations in the Negev Desert." *Nature* 332 (1988): 144–146.

Gopher, A. "Early Pottery-Bearing Groups in Israel—the Pottery Neolithic Period." Pp. 205–225 in *The Archaeology of Society in the Holy Land*, edited by T. E. Levy. London: Leicester University Press, 1998.

Gophna, R. "Early Bronze Age C Anaan: Some Spatial and Demographic Observations." Pp. 269–280 in *The Archaeology of Society in the Holy Land*, edited by T. E. Levy. London: Leicester University Press, 1998.

Gophna, R., and J. Portugali. "Settlement and Demographic Processes in Israel's Coastal Plain from the Chalcolithic to the Middle Bronze Age." *Bulletin of the American School of Oriental Research* 269 (1988): 11–28.

Goring-Morris, N. "Complex Hunter-Gatherers at the End of the Paleolithic (20,000–10,000 BP)." Pp. 141–168 in *The Archaeology of Society in the Holy Land*, edited by T. E. Levy. London: Leicester University Press, 1998.

Goring-Morris, N., and A. Belfer-Cohen. "The Articulation of Cultural Processes and Late Quaternary Environmental Changes in Cisjordan." *Paléorient* 23, no. 2 (1998): 71–93.

Greenberg, J. B., and T. K. Park. "Political Ecology." *Journal of Political Ecology* 1 (1994): 1–12.

Greene, K. "Technological Innovation and Economic Progress in the Ancient World: M. I. Finley Re-Considered." *Economic History Review* 53, no. 1 (2000): 29–59.

Gremillion, K. J. "Seed Processing and the Origins of Food Production in Eastern North America." *American Antiquity* 69, no. 2 (2004): 215–233.

Gvirtzman, G., and M. Wieder. "Climate of the Last 53,000 Years in the Eastern Mediterranean, Based on Soil-Sequence Stratigraphy in the Coastal Plain of Israel." *Quaternary Science Reviews* 20, no. 18 (2001): 1827–1849.

Haberle, S. G., and B. David. "Climates of Change: Human Dimensions of Holocene Environmental Change in Low Latitudes of the Pepii Transect." *Quaternary International* 118–119 (2004): 165–179.

Haiman, M. "Agriculture and Nomad-State Relations in the Negev Desert in the Byzantine and Early Islamic Periods." *Bulletin of the American Schools of Oriental Research* 297 (1995): 29–54.

Halstead, P., and J. O'Shea. *Bad Year Economics: Cultural Responses to Risk and Uncertainty.* Cambridge: Cambridge University Press, 1989a.

————. "Cultural Responses to Risk and Uncertainty." Pp. 1–7 in *Bad Year Economics: Cultural Responses to Risk and Uncertainty*, edited by P. Halstead and J. O'Shea. Cambridge: Cambridge University Press, 1989b.

Hammond, P. C. *The Nabataeans: Their History, Culture and Archaeology.* Gothenburg, Swed.: P. Åström, 1973.

Harris, D. R. "Climatic Change and the Beginnings of Agriculture: The Case of the Younger Dryas." Pp. 379–394 in *Evolution on Planet Earth: Impact of the Physical Environment*, edited by L. Rothschild and A. Lister. London: Academic Press, 2003.

Heim, C., N. R. Nowaczyk, J. F. W. Negendank, S. A. G. Leroy, and Z. Ben-Avraham. "Near East Desertification: Evidence from the Dead Sea." *Naturwissenschaften* 84 (1997): 398–401.

Helbaek, H. "Appendix A: Plant Economy in Ancient Lachish." Pp. 309–317 in *Lachish: Tell Ed Duweir 4, the Bronze Age*, edited by O. Tufnell. London: Oxford University Press, 1958.

Henry, D. O. *From Foraging to Agriculture: The Levant at the End of the Ice Age*. Philadelphia: University of Pennsylvania Press, 1989.

———. "Foraging, Sedentism, and Adaptive Vigor in the Natufian: Rethinking the Linkages." Pp. 353–370 in *Perspectives on the Past: Theoretical Biases in Mediterranean Hunter-Gatherer Research*, edited by G. A. Clark. Philadelphia: University of Pennsylvania Press, 1991.

Higgs, E. S., and M. R. Jarman. "The Origins of Agriculture: A Reconsideration." *Antiquity* 43 (1969): 31–41.

Hillman, G. "Late Pleistocene Changes in Wild Plant-Foods Available to Hunter-Gatherers of the Northern Fertile Crescent: Possible Preludes to Cereal Cultivation." Pp. 159–203 in *The Origins and Spread of Agriculture and Pastoralism in Eurasia*, edited by D. R. Harris. Washington, DC: Smithsonian Institution, 1996.

Hillman, G. C., and M. S. Davis. "Measured Domestication Rates in Wild Wheat and Barley under Primitive Cultivation, and Their Archaeological Implications." *Journal of World Prehistory* 4, no. 2 (1990): 157–222.

Hillman, G., R. Hedges, A. Moore, S. Colledge, and P. Pettitt. "New Evidence of Late Glacial Cereal Cultivation at Abu Hureyra on the Euphrates." *The Holocene* 11, no. 4 (2001): 383–393.

Hirschfeld, Y. "Farms and Villages in Byzantine Palestine." *Dumbarton Oaks Papers* 51 (1997): 33–71.

Holladay, J. S. "The Kingdoms of Israel and Judah: Political and Economic Centralization in the Iron IIA-B (ca. 1000–750 BCE)." Pp. 368–398 in *The Archaeology of Society in the Holy Land*, edited by T. E. Levy. London: Leicester University Press, 1998.

Holliday, V. "Quaternary Geoscience in Archaeology." Pp. 3–35 in *Earth Sciences and Archaeology*, edited by P. Goldberg, V. T. Holliday, and C. R. Ferring. London: Kluwer Press, 2001.

Hopf, M. "Appendix B: Jericho Plant Remains." Pp. 576–621 in *Excavations at Jericho V*, edited by K. M. Kenyon and T. A. Holland. London: British School of Archaeology in Jerusalem, 1983.

Horowitz, A. "Climatic and Vegetational Developments in Northeastern Israel during Upper Pleistocene-Holocene Times." *Pollen et Spores* 13 (1971): 255–278.

———. "Pollen Spectra from Two Early Holocene Prehistoric Sites in the Har Harif (West Central Negev)." Pp. 323–326 in *Prehistory and Palaeoenvironment in the Central Negev, Israel*, edited by A. E. Marks. Dallas: Southern Methodist University Press, 1976.

Hovers, E. "Settlement and Subsistence Patterns in the Lower Jordan Valley from Epipalaeolithic to Neolithic Times." Pp. 37–51 in *People and Culture in Change*, edited by I. Hershkovitz. Oxford: British Archaeological Reports, International Series, 508, 1989.

Hovers, E., and O. Bar-Yosef. "Prehistoric Survey of Eastern Samaria: A Preliminary Report." *Israel Exploration Journal* 37 (1987): 77–87.

Hunt, C. O., H. A. Elrishi, D. D. Gilbertson, J. P. Grattan, S. McLaren, F. B. Pyatt, G. Rushworth, and G. W. Barker. "Early Holocene Environments in the Wadi Faynan, Jordan." *The Holocene* 14, no. 6 (2004): 921–930.

Huntington, E. *The Pulse of Asia: A Journey in Central Asia Illustrating the Geographic Basis of History.* Boston: Houghton, Mifflin and Company, 1907.

———. *Palestine and Its Transformation.* Boston: Houghton, Mifflin and Company, 1911.

———. *Civilization and Climate.* New Haven: Yale University Press, 1924.

Ilan, D. "The Middle Bronze Age (circa 2000–1500 B.C.E.)." Pp. 331–342 in *Near Eastern Archaeology: A Reader,* edited by S. Richard. Winona Lake, IN: Eisenbrauns, 2003.

Ilan, O., and M. Sebbane. "Copper Metallurgy, Trade and the Urbanization of Southern Canaan in the Chalcolithic and Early Bronze Age." Pp. 139–162 in *L'urbanisation de la Palestine à L'âge du Bronze Ancien: Bilan et Perspectives des Recherches Actuelles: Actes du Colloque D'emmaüs (20–24 October 1986),* edited by P. de Miroschedji. Oxford: British Archaeological Reports, International Series, 527, 1989.

Ingram, M. J., D. J. Underhill, and T. M. L. Wigley. "Historical Climatology." *Nature* 276 (1978): 329–334.

Isaac, B. *The Limits of Empire: The Roman Army in the East.* Oxford, UK: Clarendon Press, 1992.

Issar, A. S. *Water Shall Flow from the Rock.* Heidelberg, Ger.: Springer-Verlag, 1990.

Issar, A. S., Y. Govrin, M. A. Geyh, E. Wakshal, and M. Wolf. "Climate Changes during the Upper Holocene in Israel." *Israel Journal of Earth Sciences* 40 (1992): 219–223.

Issar, A. S., and M. Zohar. *Climate Change: Environment and Civilization in the Middle East.* Berlin: Springer-Verlag, 2004.

Iversen, J. "The Late-Glacial Flora of Denmark and Its Relationship to Climate and Soil." *Danmarks Geologische Undersogelse Series II* 75 (1954): 1–175.

Jacobsen, T. *Toward the Image of Tammuz and Other Essays on Mesopotamian History and Culture.* Cambridge, MA: Harvard University Press, 1970.

Jacobsen, T., and R. McC. Adams. "Salt and Silt in Ancient Mesopotamian Agriculture." *Science* 128 (1958): 1251–1258.

Jensen, K. "Archaeological Dating in the History of North Jutland's Vegetation." *Acta Archaeologica* 5 (1935): 185–214.

Joffe, A. "Early Bronze I and the Evolution of Social Complexity in Canaan." *Journal of Mediterranean Archaeology* 4 (1991): 3–58.

———. *Settlement and Society in the Early Bronze Age I and II, Southern Levant.* Sheffield, UK: Sheffield Academic Press, 1993.

Jorde, L. B. "Precipitation Cycles and Cultural Buffering in the Prehistoric Southwest." Pp. 385–396 in *For Theory Building in Archaeology: Essays on Faunal Remains, Aquatic Resources, Spatial Analysis, and Systemic Modeling,* edited by L. R. Binford. New York: Academic Press, 1977.

Kallel, N., M. Paterne, L. Labeyrie, J. Duplessy, and M. Arnold. "Temperature and Salinity Records of the Tyrrhenian Sea During the Last 18,000 Years." *Palaeogeography Palaeoclimatology Palaeoecology* 135 (1997): 97–108.

Kempinski, A. *Megiddo: A City-State and Royal Centre in North Israel.* Munich: Verlag, 1989.

Kennett, D. J., and J. P. Kennett. "Competitive and Cooperative Responses to Climatic Instability in Coastal Southern California." *American Antiquity* 65, no. 2 (2000): 379–395.

Kenyon, K. M. *Excavations at Jericho I: The Tombs Excavated in 1952–1954.* London: British School of Archaeology in Jerusalem, 1960.

Khotinskiy, N. A. "Holocene Vegetation History." Pp. 179–200 in *Late Quaternary Environments of the Soviet Union, English Edition,* edited by A. Velichko, H. E. Wright, and C. W. Barnosky. Minneapolis: University of Minnesota Press, 1984.

Kislev, M. E. "Agriculture in the near East in the 7th Millennium B.C." Pp. 51–55 in *Prehistory of Agriculture: New Experimental and Ethnographic Approaches,* edited by P. Anderson. Los Angeles: University of California Press, 1999.

Kislev, M. E., D. Nadel, and I. Carmi. "Epipaleolithic (19,000 BP) Cereal and Fruit Diet at Ohalo II, Sea of Galilee, Israel." *Review of Palaeobotany and Palynology* 73 (1992): 161–166.

Klinger, Y., J. P. Avouac, D. Bourles, and N. Tisnerat. "Alluvial Deposition and Lake-Level Fluctuations Forced by Late Quaternary Climate Change: The Dead Sea Case Example." *Sedimentary Geology* 162, no. 1/2 (2003): 119–139.

Knapp, A. B. "Independence and Imperialism: Politico-Economic Structures in the Bronze Age Levant." Pp. 83–98 in *Archaeology, Annales, and Ethnohistory,* edited by A. B. Knapp. Cambridge: Cambridge University Press, 1992.

Knaut, A. "Pueblo Revolt of 1680: Eighty Years of Cultural Tension." *Artifact* 27, no. 4 (1989): 17–94.

———. *The Pueblo Revolt of 1680.* Norman: University of Oklahoma Press, 1995.

Kramer, S. N. "Dumuzi's Annual Resurrection: An Important Correction to 'Inanna's Descent.'" *Bulletin of the American School of Oriental Research* 183 (1966): 31.

Kroon, D., I. Alexander, M. Little, L. J. Lourens, A. Matthewson, A. H. F. Robertson, and T. Sakamoto. "Oxygen Isotope and Sropel Stratigraphy in the Eastern Mediterranean during the Last 3.2 Million Years." *Proceedings of the Ocean Drilling Program, Scientific Results* 160 (1998): 181–189.

Krumbein, W. C., and L. L. Sloss. *Stratigraphy and Sedimentation.* San Francisco: Freeman, 1963.

Kuijt, I., and N. Goring-Morris. "Foraging, Farming, and Social Complexity in the Pre-Pottery Neolithic of the Southern Levant: A Review and Synthesis." *Journal of World Prehistory* 16, no. 4 (2002): 361–440.

Kuzucuoğlu, C., and N. Roberts. "Évolution de L'environnment en Anatolie de 20,000 à 6,000 BP." *Paléorient* 23, no. 2 (1998): 7–24.

LaBianca, Ø. S., and R. W. Younker. "The Kingdoms of Ammon, Moab and Edom: The Archaeology of Society in Late Bronze/Iron Age Transjordan (ca. 1400–500 B.C.E.)." Pp. 399–415 in *The Archaeology of Society in the Holy Land,* edited by T. E. Levy. London: Leicester University Press, 1998.

Lamb, H. H. *Climate, History and the Modern World.* London: Routledge

Lartet, L. "Sur la Formation du Bassin de la Mer Morte ou Lac Asr_ Survenus dans le Niveau de ce Lac." *Comptes Rend_ (1865): 796–800.

_, 1995.
_naltite, et sur les Changements
_Académie des Sciences, Paris 60
_s de l'Académie

Lauritzen, S. E. "High-Resolution Paleotemperature Proxy Record for the Last Interglaciation Based on Norwegian Speleothems." *Quaternary Research* 43 (1995): 133–146.

Lavee, H., J. Poesen, and A. Yair. "Evidence of High Efficiency Water-Harvesting by Ancient Farmers in the Negev Desert, Israel." *Journal of Arid Environments* 35, no. 2 (1997): 341–348.

Lemcke, G., and M. Sturm. "$\delta^{18}O$ and Trace Element Measurements as Proxy for the Reconstruction of Climate Changes at Lake Van (Turkey): Preliminary Results." Pp. 653–678 in *Third Millennium BC Climate Change and Old World Collapse*, edited by H. N. Dalfes, G. Kukla, and H. Weiss. Berlin: Springer-Verlag, 1997.

Lender, Y. *Archaeological Survey of Israel Map of Har Nafha (196) 12-01*. Jerusalem: Israel Antiquities Authority, 1990.

Leroi-Gourhan, A., and F. Darmon. "Analyses Polliniques de Stations Natoufiennes au Proche Orient." Pp. 21–26 in *The Natufian Culture in the Levant*, edited by O. Bar-Yosef and F. R. Valla. Ann Arbor, MI: International Monographs in Prehistory, 1991.

Levy, J. P. *The Economic Life of the Ancient World*. Chicago: University of Chicago Press, 1967.

Levy, T. E. "Transhumance, Subsistence, and Social Evolution." Pp. 65–82 in *Pastoralism in the Levant: Archaeological Materials in Anthropological Perspectives*, edited by O. Bar-Yosef and A. M. Khazanov. Madison, WI: Prehistory Press, 1992.

———. "Cult, Metallurgy and Rank Societies: Chalcolithic Period (ca. 4500–3500 BCE)." Pp. 227–244 in *The Archaeology of Society in the Holy Land*, edited by T. E. Levy. London: Leicester University Press, 1998a.

———. "Preface." Pp. x–xvi in *The Archaeology of Society in the Holy Land*, edited by T. E. Levy. London: Leicester University Press, 1998b.

Lieberman, D. E., T. W. Deacon, and R. H. Meadow. "Computer Image Enhancement and Analysis of Cementum Increments as Applied to Teeth of *Gazella gazella*." *Journal of Archaeological Science* 17 (1990): 519–533.

Liphschitz, N. "The Vegetational Landscape and the Macroclimate of Israel during Prehistoric and Protohistoric Periods." *Mitkufat Haeven, N.S.* 19 (1986): 80–90.

———. "Plant Economy and Diet in the Early Bronze Age in Israel: A Summary of Present Research." Pp. 269–277 in *L'urbanisation de la Palestine à L'âge du Bronze Ancien: Bilan et Perspectives des Recherches Actuelles: Actes du Colloque D'emmaüs (20–24 October 1986)*, edited by P. de Miroschedji. Oxford: British Archaeological Reports, International Series, 527, 1989.

Liphschitz, N., R. Gophna, and S. Lev-Yadun. "Man's Impact on the Vegetational Landscape of Israel in the Early Bronze Age Ii–Iii." Pp. 263–267 in *L'urbanisation de la Palestine à L'âge du Bronze Ancien: Bilan et Perspectives des Recherches Actuelles: Actes du Colloque D'emmaüs (20–24 October 1986)*, edited by P. de Miroschedji. Oxford: British Archaeological Reports, International Series, 527, 1989.

Low, Bobbi S. "Human Responses to Environmental Extremeness and Uncertainty: A Cross-Cultural Perspective." Pp. 229–255 in *Risk and Uncertainty in Tribal and Peasant Economies*, edited by E. Cashdan. Boulder, CO: Westview, 1990.

Lowe, J. J., and M. J. C. Walker. *Reconstructing Quaternary Environments*. Essex, UK: Addison, 1997.

Mackay, A. "Climate and Popular Unrest in Late Medieval Castile." Pp. 356–375 in *Climate and History: Studies in Past Climates and Their Impact on Man*, edited by T. M. L. Wigley, M. J. Ingram, and G. Farmer. Cambridge: Cambridge University Press, 1981.

Macklin, M. G., J. Lewin, and J. C. Woodward. "Quaternary Fluvial Systems in the Mediterranean Basin." Pp. 1–25 in *Mediterranean Quaternary River Environments*, edited by J. Lewin, M. G. Macklin, and J. C. Woodward. Rotterdam: Balkema, 1995.

Maisler, B., M. Stekelis, and M. Avi-Yonah. "The Excavations at Beth Yerah (Khirbet Kerak) 1944–1946." *Israel Exploration Journal* 2 (1952): 218–229.

Mayerson, P. "The Ancient Agricultural Regime of Nessana and the Central Negeb." *Excavations at Nessana: (Auja Hafir, Palestine) Vol. 1*, edited by H. D. Colt. London: British School of Archaeology in Jerusalem, 1962.

———. "A Note on Demography and Land Use in the Ancient Negeb." *Bulletin of the American Schools of Oriental Research* 185 (1967): 39–43.

———. "Wine and Vineyards of Gaza in the Byzantine Period." *Bulletin of the American School of Oriental Research* 257 (1985): 75–80.

———. "Saracens and Roman: Micro-Macro Relationships." *Bulletin of the American Schools of Oriental Research* 274 (1989): 71–79.

Mazor, G. "The Wine Presses of the Negev." *Qadmoniot* 14 (1981): 51–60 (in Hebrew).

McClure, H. A. "Radiocarbon Chronology of Late Quaternary Lakes in the Arabian Desert." *Nature* 263 (1976): 755–756.

McCorriston, J. "Acorn Eating and Agricultural Origins: California Ethnographies as Analogies for the Ancient near East." *Antiquity* 68, no. 258 (1994): 97–107.

McCorriston, J., and F. Hole. "The Ecology of Seasonal Stress and the Origins of Agriculture in the Near East." *American Anthropologist* 93 (1991): 46–69.

McGovern, T. H. "Management for Extinction in Norse Greenland." Pp. 127–154 in *Historical Ecology*, edited by C. Crumley. Santa Fe, NM: School of American Research, 1994.

McIntosh, R. J., J. A. Tainter, and S. K. McIntosh. "Climate, History, and Human Action." Pp. 1–42 in *The Way the Wind Blows: Climate, History, and Human Action*, edited by R. J. McIntosh, J. A. Tainter, and S. K. McIntosh. New York: Columbia University Press, 2000a.

———. *The Way the Wind Blows: Climate, History, and Human Action*. New York: Columbia University Press, 2000b.

McLaren, S. J., D. D. Gilbertson, J. P. Grattan, C. O. Hunt, G. A. T. Duller, and G. A. Barker. "Quaternary Palaeogeomorphologic Evolution of the Wadi Faynan Area, Southern Jordan." *Palaeogeography, Palaeoclimatology, Palaeoecology* 205, no. 1–2 (2004): 131–154.

Meadows, J. "The Younger Dryas Episode and the Radiocarbon Chronologies of the Lake Huleh and Ghab Valley Pollen Diagrams, Israel and Syria." *The Holocene* 15, no. 4 (2005): 631–636.

Meltzer, D. J. "Human Responses to Middle Holocene (Altithermal) Climates on the North American Great Plains." *Quaternary Research* 52, no. 3 (1999): 404–416.

Minnis, P. E. *Social Adaptation to Food Stress: A Prehistoric Southwestern Example*. Chicago: University of Chicago Press, 1985.

de Miroschedji, P. *Rapport sur les Trois Premières Campagnes de Fouilles à Tel Yarmouth, (Israël): (1980–1982)*. Paris: Editions Recherche sur les Civilisations, 1988.

Moore, A. M. T., and G. C. Hillman. "The Pleistocene to Holocene Transition and Human Economy in Southwest Asia: The Impact of the Younger Dryas." *American Antiquity* 57, no. 3 (1992): 482–494.

Moore, P. E., J. A. Webb, and M. D. Collinson. *Pollen Analysis*. Oxford, UK: Blackwell, 1991.

Munro, N. D. "Zooarchaeological Measures of Hunting Pressure and Occupation Intensity in the Natufian: Implications for Agricultural Origins." *Current Anthropology* 45 (2004): S5–S33.

Neev, D., and K. O. Emery. *The Dead Sea: Depositional Processes and Environments of Evaporites*. Jerusalem: Monson, 1967.

———. *The Destruction of Sodom, Gomorrah, and Jericho: Geological Climatological, and Archaeological Background*. Oxford: Oxford University Press, 1995.

Neev, D., and J. Hall. "Climatic Fluctuations during the Holocene as Reflected by the Dead Sea Levels." Pp. 53–60 in *Desertic Terminal Lakes*, edited by D. Greer. Logan: Utah State University, 1977.

Negev, A. *The Architecture of Mampsis. Part I: The Middle and Late Nabatean Periods*. Jerusalem: Institute of Archaeology, Hebrew University, 1988a.

———. *The Architecture of Mampsis Final Report Volume 2: The Late Roman and Byzantine Periods*. Jerusalem: Institute of Archaeology, Hebrew University, 1988b.

———. *Personal Names in the Nabatean Realm*. Jerusalem: Institute of Archaeology, Hebrew University, 1991.

Nesbitt, M. "When and Where Did Domesticated Cereals First Occur in Southwest Asia?" Pp. 113–132 in *The Dawn of Farming in the Near East*, edited by R. T. J. Cappers and S. Bottema. Berlin: Ex Oriente, 2002.

Niklewski, J., and W. van Zeist. "A Late Quaternary Pollen Diagram from Northwestern Syria." *Acta Botanica Neerlandica* 9, no. 5 (1970): 737–754.

Nilsson, S., and D. Pitt. *Protecting the Atmosphere: The Climate Change Convention and Its Context*. London: Earthscan Publications, 1994.

Orni, E., and E. Efrat. *Geography of Israel*. Jerusalem: Israel Universities Press, 1980.

Palmer, E. H. *The Desert of the Exodus: Journeys on Foot in the Wilderness of the Forty Years' Wanderings*. Cambridge, UK: Deighton, Bell and Co., 1871.

Parker, S. T. "Peasants, Pastoralists, and the *Pax Romana*: A Different View." *Bulletin of the American Schools of Oriental Research* 265 (1987): 35–51.

———. "An Empire's New Holy Land: The Byzantine Period." *Near Eastern Archaeology* 62, no. 3 (1999): 134–180.

Peet, R., and M. Watts. "A Political Ecology for the 1990's." *Economic Geography* 69, no. 3 (1994): 238–242.

Perrot, J. "Le Gisement Natoufien de Mallaha (Eynan), Israël." *L'Anthropologie* 70 (1966): 437–484.

Piperno, D. R., E. Weiss, I. Holst, and D. Nadel. "Processing of Wild Cereal Grains in the Upper Palaeolithic Revealed by Starch Grain Analysis." *Nature* 430 (2004): 670–673.

Pisias, N. G., D. G. Martinson, Jr., T. C. Moore, N. J. Shackleton, W. Prell, J. Hays, and G. Boden. "High Resolution Stratigraphic Correlation of Benthic Oxygen Isotopic Records Spanning the Last 300,000 Years." *Marine Geology* 56 (1984): 119–136.

Prag, K. "Preliminary Report on the Excavations at Tell Iktanu, Jordan, 1987." *Levant* XXI (1989): 33–45.

Pumpelly, R., ed. *Explorations in Turkestan: Expedition of 1904: Prehistoric Civilizations of Anau: Origins, Growth, and Influence of Environment.* Washington, DC: Carnegie Institution of Washington, 1908.

Purseglove, J. W. *Tropical Crops: Monocotyledons.* New York: Halsted, 1972.

Randsborg, K. *The First Millennium A.D. in Europe and the Mediterranean: An Archaeological Essay.* Cambridge: Cambridge University Press, 1991.

Rao, N. V. K. "Impact of Drought on the Social System of a Telengana Village." *The Eastern Anthropologist* 27, no. 4 (1974): 299–314.

Rappaport, R. A. "Maladaptation in Social Systems." Pp. 49–71 in *The Evolution of Social Systems*, edited by J. Friedman and M. J. Rowlands. London: Duckworth, 1977.

Rast, W., and T. Schaub, eds. *The Southeastern Dead Sea Plain Expedition: An Interim Report of the 1977 Season.* Cambridge, MA: American Schools of Oriental Research, 1981.

Rathbone, D. "The Grain Trade and Grain Shortages in the Hellenistic East." Pp. 45–55 in *Trade and Famine in Classical Antiquity*, edited by P. Garnsey and C. R. Whittaker. Cambridge, UK: Cambridge Philological Society, 1983.

Ravikovitch, S. *Manual and Map of Soils of Israel.* Jerusalem: Magnes, Hebrew University, 1969.

Reff, D. T. "The 'Predicament of Culture' and Spanish Missionary Accounts of the Tepehuan and Pueblo Revolts." *Ethnohistory* 42, no. 1 (1995): 63–90.

Renfrew, J. M. *Palaeoethnobotany: The Prehistoric Food Plants of the Near East and Europe.* New York: Columbia University Press, 1973.

Richard, S. "Expedition to Khirbet Iskander and Its Vicinity: Fourth Preliminary Report." *Bulletin of the American Schools of Oriental Research, Supplement* 26 (1990): 33–58.

———. "The Early Bronze Age in the Southern Levant." Pp. 286–302 in *Near Eastern Archaeology: A Reader*, edited by S. Richard. Winona Lake, IN: Eisenbrauns, 2003.

Rindos, D. *The Origins of Agriculture: An Evolutionary Perspective.* New York: Academic Press, 1984.

Robbins, P. *Political Ecology: A Critical Introduction.* Oxford, UK: Blackwell, 2004.

Roberts, N. "Late Quaternary Geomorphological Change and the Origins of Agriculture in South Central Turkey." *Geoarchaeology* 6, no. 1 (1991): 1–26.

———. *The Holocene: An Environmental History.* Oxford, UK: Blackwell, 1998.

———. "Did Prehistoric Landscape Management Retard the Post-Glacial Spread of Woodland in Southwest Asia?" *Antiquity* 76 (2002): 1002–1010.

Roberts, N., C. Kuzucuoğlu, and M. Karabıyıkoğlu, eds. *The Late Quaternary in the Eastern Mediterranean Region.* Oxford, UK: Pergamon, 1999.

Roberts, N., M. Meadows, and J. Dodson. "The History of Mediterranean-Type Environments: Climate, Culture and Landscape." *The Holocene* 11 (2001): 631–634.

Robinson, G. *Voyage en Palestine et en Syrie.* Paris: A. Bertrand, 1837.

Rohling, E. J., F. J. Jorissen, and H. C. De Stigter. "200 Year Interruption of Holocene Sapropel Formation in the Adriatic Sea." *Journal of Micropalaeontology* 16, no. 2 (1997): 97–108.

Rosen, A. M. *Cities of Clay: The Geoarcheology of Tells.* Chicago: University of Chicago Press, 1986a.

———. "Environmental Change and Settlement at Tel Lachish, Israel." *Bulletin of the American Schools of Oriental Research* 263 (1986b): 55–60.

———. *Quaternary Alluvial Stratigraphy of the Shephela and Its Paleoclimatic Implications.* Geological Survey of Israel Report, GSI/25/86. Jerusalem: Geological Survey of Israel, 1986c.

———. "Environmental Change at the End of Early Bronze Age Palestine." Pp. 247–255 in *L'Urbanisation de la Palestine à l'Age du Bronze Ancien,* edited by P. de Miroschedji. Oxford: British Archaeological Reports, 1989.

———. "Early Bronze Age Tel Erani: An Environmental Perspective." *Tel Aviv* 18 (1991): 192–204.

———. "The Social Response to Environmental Change in Early Bronze Age Canaan." *Journal of Anthropological Archaeology* 14 (1995): 26–44.

———. "The Agricultural Base of Urbanism in the Early Bronze II–III Levant." Pp. 92–98 in *Urbanism in Antiquity: From Mesopotamia to Crete,* edited by W. E. Aufrecht, N. A. Mirau and S. W. Gauley. Sheffield, UK: Sheffield Academic Press, 1997a.

———. "Environmental Change and Human Adaptational Failure at the End of the Early Bronze Age in the Southern Levant." Pp. 25–38 in *Third Millennium BC Climate Change and Old World Collapse,* edited by H. N. Dalfes, G. Kukla, and H. Weiss. Heidelberg, Ger.: Springer-Verlag, 1997b.

———. "The Geoarchaeology of Holocene Environments and Land Use at Kazane Höyük, S. E. Turkey." *Geoarchaeology* 12 (1997c): 395–416.

———. "Early to Mid-Holocene Environmental Changes and Their Impact on Human Communities in Southeastern Anatolia." Pp. 215–240 in *Water, Environment and Society in Times of Climatic Change,* edited by A. Issar, N. Brown, and Z. Shilony. New York: Kluwer Press, 1998.

———. "Phytolith Analysis in Near Eastern Archaeology." Pp. 86–92 in *The Practical Impact of Science on Aegean and Near Eastern Archaeology,* edited by S. Pike and S. Gitin. London: Archetype Press, 1999.

———. "Environmental Studies." Pp. 153–156 in *Tel Te'o, a Neolithic, Chalcolithic, and Early Bronze Age Site in the Hula Valley,* edited by E. Eisenberg, A. Gopher, and R. Greenberg. Jerusalem: Israel Antiquities Authority, 2001.

———. "Climate Change, Landscape, and Shifting Agricultural Potential during the Occupation of Tel Megiddo." Pp. 441–449 in *Megiddo IV,* edited by I. Finkelstein, D. Ussishkin, and B. Halpern. Winona Lake, IN: Eisenbrauns, 2006.

———. "Geomorphological Setting and Paleoenvironments of the Pottery Neolithic Site at Tel Yosef." *Excavations at Tel Yosef,* edited by K. Covello-Paran. Jerusalem: Israel Antiquity Authority, in press.

———. "Preliminary Geological Observations of Early Bronze Age Landforms at Har Tuv." Unpublished Manuscript on file at the Institute of Archaeology, University College, London.

Rosen, A. M., C. Chang, and F. Pavlovich Grigoriev. "Paleoenvironments and Economy of Iron Age Saka-Wusun Agropastoralists in Southeastern Kazakhstan." *Antiquity* 74 (2000): 611–623.

Rosen, A. M., and S. A. Rosen. "Determinist or Not Determinist? Climate, Environment, and Archaeological Explanation in the Levant." Pp. 535–549 in *Studies in the Archaeology of Israel and Neighboring Lands in Memory of Douglas L. Esse,* edited by S. R. Wolff. Chicago: Oriental Institute, University of Chicago, 2001.

Rosen, A. M., and S. Weiner. "Identifying Ancient Irrigation: A New Method Using Opaline Phytoliths from Emmer Wheat." *Journal of Archaeological Science* 21 (1994): 132–135.

Rosen, S. A. "Demographic Trends in the Negev Highlands: Preliminary Results from the Emergency Survey." *Bulletin of the American Schools for Oriental Research* 266 (1987): 45–58.

———. "Craft Specialization and the Rise of Secondary Urbanism: A View from the Southern Levant." Pp. 82–91 in *Urbanism in Antiquity: From Mesopotamia to Crete*, edited by W. E. Aufrecht, N. A. Mirau, and S. W. Gauley. Sheffield, UK: Sheffield Academic Press, 1997.

———. "The Decline of Desert Agriculture: A View from the Classical Period Negev." Pp. 45–62 in *The Archaeology of Drylands: Living at the Margin*, edited by G. Barker and D. D. Gilbertson. London: Routledge, 2000.

———. Climatic Determinism and Social Collapse: A Critique from Archaeology. Paper presented at the Annual Earth Sciences Day of the Israel Geological Society, Weizmann Institute, Rehovot, Israel, 2004.

Rossignol-Strick, M. "Sea-Land Correlation of Pollen Records in the Eastern Mediterranean for the Glacial-Interglacial Transition: Biostratigraphy versus Radiometric Time-Scale." *Quaternary Science Reviews* 14 (1995): 893–915.

———. "Paléoclimat de la Méditerranée Orientale et de l'Asie du Sud-Ouest de 15,000 à 6,000 BP." *Paléorient* 23, no. 2 (1997): 175–186.

———. "The Holocene Climatic Optimum and Pollen Records of Sapropel 1 in the Eastern Mediterranean, 9000–6000 BP." *Quaternary Science Reviews* 18, no. 4–5 (1999): 515–530.

Rowley-Conwy, P., and M. Zvelebil. "Saving It for Later: Storage by Prehistoric Hunter-Gatherers in Europe." Pp. 40–56 in *Bad Year Economics: Cultural Responses to Risk and Uncertainty*, edited by P. Halstead and J. O'Shea. Cambridge: Cambridge University Press, 1989.

Rubin, R. "The Debate over Climate Changes in the Negev 4th–7th Centuries C.E., Palestine." *Palestine Exploration Quarterly* (1989): 71–78.

———. "Settlement and Agriculture on an Ancient Desert Frontier." *The Geographical Review* 81, no. 2 (1991): 197–205.

———. "Urbanization, Settlement, and Agriculture in the Negev Desert—the Impact of the Roman-Byzantine Empire on the Frontier." *Zeitschrift des Deutschen Palästini-Vereins* 112 (1996): 49–60.

———. "The Romanization of the Negev, Israel: Geographical and Cultural Changes in the Desert Frontier in Late Antiquity." *Journal of Historical Geography* 23, no. 3 (1997): 267–283.

Sanlaville, P. "L'Espace Géographique de Mari." *MARI* 4 (1985): 15–26.

Schilman, B., A. Ayalon, M. Bar-Matthews, E. J. Kagan, and A. Almogi-Labin. "Sea-Land Paleoclimate Correlation in the Eastern Mediterranean Region during the Late Holocene." *Israel Journal of Earth Sciences* 51 (2002): 181–190.

Schilman, B., M. Bar-Matthews, A. Almogi-Labin, and B. Luz. "Global Climate Instability Reflected by Eastern Mediterranean Marine Records during the Late Holocene." *Palaeogeography Palaeoclimatology Palaeoecology* 176, no. 1–4 (2001): 157–176.

Schuldenrein, J., and G. A. Clark. "Prehistoric Landscapes and Settlement Geography along the Wadi Hasa, West-Central Jordan. Part I: Geoarchaeology, Human Palaeoecology and Ethnographic Modelling." *Environmental Archaeology* 6 (2001): 23–38.

Schuldenrein, J., and P. Goldberg. "Late Quaternary Palaeoenvironments and Prehistoric Site Distribution in the Lower Jordan Valley: A Preliminary Report." *Paléorient* 7 (1981): 57–72.

Schwab, M. J., F. Neumann, T. Litt, J. Negendank, and M. Stein. "Holocene Palaeoecology of the Golan Heights (Near East): Investigation of Lacustrine Sediments from Birket Ram Crater Lake." *Quaternary Science Reviews* 23, no. 16–17 (2004): 1723–1732.

Scott, L. "The Holocene of Middle Latitude Arid Areas." Pp. 396–405 in *Global Change in the Holocene*, edited by A. Mackay, R. Battarbee, J. Birks, and F. Oldfield. London: Arnold, 2003.

Seger, J. D. "Some Provisional Correlations in EB III Stratigraphy in Southern Palestine." Pp. 117–135 in *L'urbanisation de la Palestine à L'âge du Bronze Ancien: Bilan et Perspectives des Recherches Actuelles: Actes du Colloque D'emmaüs (20–24 October 1986)*, edited by P. de Miroschedji. Oxford: British Archaeological Reports, International Series, 527, 1989.

Shackleton, N. J. "Oxygen Isotope Analyses and Pleistocene Temperatures Re-Assessed." *Nature* 215 (1967): 15–17.

Shereshevski, J. *Byzantine Urban Settlements in the Negev Desert*. Beer Sheva: Ben-Gurion University of the Negev Press, 1991.

Sherratt, A. G. "Water, Soil and Seasonality in Early Cereal Cultivation." *World Archaeology* 11 (1980): 313–330.

———. "Plough and Pastoralism: Aspects of the Secondary Products Revolution." Pp. 261–305 in *Patterns of the Past: Studies in Honour of David Clarke*, edited by I. Hodder, G. L. Isaac, and N. Hammond. Cambridge: Cambridge University Press, 1981.

Silverberg, R. *The Pueblo Revolt*. New York: Weybright and Talley, 1970.

Sinopoli, C. "The Archaeology of Empires." *Annual Review of Anthropology* 23 (1994): 159–180.

Slobodkin, L. B., and A. Rapoport. "An Optimal Strategy of Evolution." *The Quarterly Review of Biology* 49 (1974): 181–200.

Smith, M. *The Ugaritic Baal Cycle: Introduction with Text, Translation and Commentary of Ktu 1.1–1.2*. Leiden, Neth.: Brill, 1994.

Sperber, D. *Roman Palestine: 200–400, the Land*. Ramat-Gan, Isr.: Bar-Ilan University, 1978.

Stager, L. "The First Fruits of Civilization." Pp. 172–188 in *Palestine in the Bronze and Iron Ages: Papers in Honor of Olga Tufnell*, edited by J. N. Tubb. London: Institute of Archaeology, 1985.

———. "The Impact of the Sea Peoples (1185–1050 BCE)." Pp. 332–348 in *The Archaeology of Society in the Holy Land*, edited by T. E. Levy. London: Leicester University Press, 1998.

Stern, E. "Between Persia and Greece: Trade, Administration and Warfare in the Persian and Hellenistic Periods (539–63 BCE)." Pp. 432–445 in *The Archaeology of Society in the Holy Land*, edited by T. E. Levy. London: Leicester University Press, 1998.

Stordeur, D. "La Contribution de l'Industrie de L'os à la Délimitation des Aires Culturelles: L'example du Natoufien." Pp. 433–437 in *Prehistoire du Levant*, edited by J. Cauvin and P. Sanlaville. Paris: CNRS, 1981.

Street-Perrott, F. A., and R. A. Perrott. "Holocene Vegetation, Lake-Levels and Climate of Africa." Pp. 318–356 in *Global Climates since the Last Glacial Maximum*, edited by H. E. J. Wright, J. E. Kutzbach, T. Webb, W. E. Ruddiman, F. A. Street-Perrott and P. J. Bartlein. Minneapolis: University of Minnesota Press, 1993.

Tainter, J. A. *The Collapse of Complex Societies.* Cambridge: Cambridge University Press, 1988.

———. "Sustainability of Complex Societies." *Futures* 27 (1995): 397–407.

Tchernov, E. "Biological Evidence for Human Sedentism in Southwest Asia during the Natufian." Pp. 315–340 in *The Natufian Culture in the Levant,* edited by O. Bar-Yosef and F. R. Valla. Ann Arbor, MI: International Monographs in Prehistory, 1991.

Thompson, R., G. M. Turner, M. Stiller, and A. Kaufman. "Palaeoclimatic Secular Variations Recorded in Sediments from the Sea of Galilee (Lake Kinneret)." *Quaternary Research* 23 (1985): 175–188.

Troen, I. "Calculating the Economic Absorptive Capacity of Palestine: A Study of the Political Uses of Scientific Research." *Contemporary Jewry* 10 (1989): 19–38.

Turner, M., and B. O'Connell. *The Whole World's Watching: Decarbonizing the Economy and Saving the World.* Chichester, UK: John Wiley, 2001.

Valla, F. "Les Etablissements Natoufiens dans le Nord d'Israel." Pp. 409–420 in *Préhistoire du Levant,* edited by J. Cauvin and P. Sanlaville. Paris: CNRS, 1981.

———. "The First Settled Societies: Natufian (12,500–10,200 BP)." Pp. 169–187 in *The Archaeology of Society in the Holy Land,* edited by T. E. Levy. London: Leicester University Press, 1998a.

———. "Natufian Seasonality: A Guess." Pp. 93–108 in *Seasonality and Sedentism: Archaeological Perspectives from Old and New World Sites,* edited by T. R. Rocek and O. Bar-Yosef. Cambridge, MA: Peabody Museum of Archaeology and Ethnology, 1998b.

Valla, F. R., H. Khalaily, H. Valladas, N. Tisnérat-LaBorde, N. Samuelian, F. Bocquentin, R. Rabinovich, A. Bridault, T. Simmons, G. Le Dosseur, A. M. Rosen, L. Dubreuil, and D. E. Bar-Yosef Mayer. "Les Fouilles de Mallaha en 2000 et 2001: 3ème Rapport Préliminaire." *Journal of the Israel Prehistoric Society* 34 (2004): 49–244.

van Geel, B., N. A. Bokovenko, N. D. Burova, K. V. Chugunov, V. A. Dergachev, V. G. Dirksen, M. Kulkova, A. Nagler, H. Parzinger, and J. van der Plicht. "Climate Change and the Expansion of the Scythian Culture after 850 BC: A Hypothesis." *Journal of Archaeological Science* 31, no. 12 (2004): 1735.

van Zeist, W., and S. Bottema. "Vegetational History of the Eastern Mediterranean and the Near East during the Last 20,000 Years." Pp. 277–323 in *Palaeoclimates, Palaeoenvironments and Human Communities in the Eastern Mediterranean Region in Later Prehistory,* edited by J. L. Bintliff and W. van Zeist. Oxford: British Archaeological Reports, International Series 133, 1982.

van Zeist, W., and H. Woldring. "A Postglacial Pollen Diagram from Lake Van in East Anatolia." *Review of Palaeobotany and Palynology* 26 (1978): 249–276.

Vita-Finzi, C. *The Mediterranean Valleys.* Cambridge: Cambridge University Press, 1969.

Wagstaff, J. M. "Buried Assumptions: Some Problems in the Interpretation of the 'Younger Fill' Raised by Recent Data from Greece." *Journal of Archaeological Science* 8 (1981): 247–264.

Weinstein-Evron, M. "Palynological History of the Last Pleniglacial in the Levant." Pp. 9–25 in *Les Industries à Pointes Foliacées du Paléolithique Supérieur Européen, Kraków 1989,* edited by D. Vialou and A. Vilhena-Vialou. Liège, Belg.: E.R.A.U.L., 1990.

Weiss, E., W. Wetterstrom, D. Nadel, and O. Bar-Yosef. "The Broad Spectrum Revisited: Evidence from Plant Remains." *Proceedings of the National Academy of Sciences of the United States of America* 101, no. 26 (2004): 9551–9555.

Weiss, H., and R. S. Bradley. "What Drives Societal Collapse?" *Science* 291, no. 5504 (2001): 609–610.

Weiss, H., M.-A. Courty, W. Wetterstrom, F. Guichard, L. Senior, R. Meadow, and A. Curnow. "The Genesis and Collapse of Third Millennium North Mesopotamian Civilization." *Science* 261 (1993): 995–1004.

Whittaker, C. R. "Trade and the Aristocracy in the Roman Empire." *Opus, International Journal for Social and Economic History of Antiquity* 4 (1988): 49–75.

Wick, L., G. Lemcke, and M. Sturm. "Evidence of Late Glacial and Holocene Climatic Change and Human Impact in Eastern Anatolia: High Resolution Pollen, Charcoal, Isotopic and Geochemical Records from the Laminated Sediments of Lake Van, Turkey." *The Holocene* 13, no. 5 (2003): 665–675.

Wilkinson, T. J. "Water and Human Settlement in the Balikh Valley, Syria: Investigations from 1992–1995." *Journal of Field Archaeology* 25, no. 1 (1998): 63–87.

———. "Holocene Valley Fills of Southern Turkey and Northwestern Syria: Recent Geoarchaeological Contributions." *Quaternary Science Reviews* 18, no. 4–5 (1999): 555–571.

———. *Archaeological Landscapes of the Near East.* Tucson: University of Arizona Press, 2003.

Willcox, G. "Measuring Grain Size and Identifying Near Eastern Cereal Domestication: Evidence from the Euphrates Valley." *Journal of Archaeological Science* 31 (2004): 145–150.

Williams, M., D. Dunkerley, P. De Deckker, A. Kershaw, and T. Stokes. *Quaternary Environments.* New York: Arnold, 1998.

Winograd, I. J., J. M. Landwehr, K. R. Ludwig, T. B. Coplen, and A. C. Rigg. "Duration and Structure of the Past Four Interglaciations." *Quaternary Research* 48 (1997): 141–154.

Winterhalder, B. "Open Field, Common Pot: Harvest Variability and Risk Avoidance in Agricultural and Foraging Societies." Pp. 67–87 in *Risk and Uncertainty in Tribal and Peasant Economies,* edited by E. Cashdan. Boulder, CO: Westview, 1990.

Winterhalder, B., and C. Goland. "An Evolutionary Ecology Perspective on Diet Choice, Risk and Plant Domestication." Pp. 123–160 in *People, Plants, and Landscapes: Studies in Paleoethnobotany,* edited by K. J. Gremillion. Tuscaloosa: University of Alabama Press, 1997.

Woolf, G. "Imperialism, Empire and the Integration of the Roman Economy." *World Archaeology* 23, no. 3 (1992): 283–293.

Woolley, C. L, and T. E. Lawrence. *The Wilderness of Zin.* London: Palestine Exploration Fund, 1914.

Wright, H. E. "Climate and Prehistoric Man in the Eastern Mediterranean." Pp. 71–97 in *Prehistoric Investigations in Iraqi Kurdistan,* edited by R. J. Braidwood and B. Howe. Chicago: Oriental Institute, University of Chicago, 1960.

———. "Environmental Determinism in Near Eastern Prehistory." *Current Anthropology* 34, no. 4 (1993): 458–469.

Wright, H. E., Jr., and J. L. Thorpe. "Climatic Change and the Origin of Agriculture in the Near East." Pp. 49–62 in *Global Change in the Holocene,* edited by A. Mackay, R. Battarbee, J. Birks, and F. Oldfield. London: Arnold, 2003.

Wright, K. I. "Ground-Stone Tools and Hunter-Gatherer Subsistence in Southwest Asia—Implications for the Transition to Farming." *American Antiquity* 59, no. 2 (1994): 238–263.

Yasuda, Y., H. Kitagawa, and T. Nakagawa. "The Earliest Record of Major Anthropogenic Deforestation in the Ghab Valley, Northwest Syria: A Palynological Study." *Quaternary International*
73/74 (2000): 127–136.

Yechieli, Y., M. Magaritz, Y. Levy, U. Weber, U. Kafri, W. Woelfli, and G. Bonani. "Late Quaternary
Geological History of the Dead Sea Area, Israel." *Quaternary Research* 39, no. 1 (1993): 59–67.

Younker, R. W. "The Iron Age in the Southern Levant." Pp. 367–382 in *Near Eastern Archaeology:
A Reader*, edited by S. Richard. Winona Lake, IN: Eisenbrauns, 2003.

Zohary, D. "Domestication of the Southwest Asian Crop Assemblage of Cereals, Pulses and Flax:
The Evidence from the Living Plants." Pp. 359–373 in *Foraging and Farming: The Evolution of
Plant Exploitation*, edited by D. R. Harris and G. Hillman. London: Unwin & Hyman, 1989.

———. "Domestication of the Neolithic Near Eastern Crop Assemblage." Pp. 51–55 in *Prehistory
of Agriculture: New Experimental and Ethnographic Approaches*, edited by P. C. Anderson. Los
Angeles: University of California Press, 1999.

Zohary, D., and M. Hopf. *Domestication of Plants in the Old World: The Origin and Spread of Cultivated Plants in West Asia, Europe, and the Nile Valley*. Oxford: Oxford University Press, 2000.

Zohary, M. *Geobotanical Foundations of the Middle East*. Stuttgart: Fischer, 1973.

———. *Vegetal Landscapes of Israel*. Tel Aviv: Am-Oved, 1980.

Index

human responses: complexity of, 102, 175; to environmental shifts, 2, 8–14, 116–17, 121–23, 146–49, 173, 179–80; lack of, leading to collapse, 177. *See also* buffering strategies

hunter-gatherers, 106; buffering strategies, 116–17; Californians as, 118–19; comparisons, 122–23; forest resources of, 122–24; in Great Plains, 120–21

Huntington, Ellsworth, 1, 103, 151

Iktanu, 144–45, 149
interglacial/glacial episodes in Pleistocene, 44–45
Inuit responses to stress, 11, 148
Iron Age, 12–13, 41–43, 91–92, 143–44
irrigation systems, 146
Iskander, 144–45, 149
isotopic data. *See* carbon isotopic data; oxygen isotopic data

Judean Hills speleothems, 48, 83–84

Kebaran hunter-gatherers, 174
Kinneret, Lake, 84–85, 91–92, 94–95

lacustrine evidence: Early Holocene, 77–78; Late Holocene, 92–94; Late Pleistocene, 64–67; Middle Holocene, 85–86
lakes. *See* Ghab, Lake; Hula, Lake; Kinneret, Lake; Lisan, Lake
landform analyses. *See* geomorphological evidence
landowner responses to stress, 9
large-scale variability, 4–6
Late Bronze Age (LB), 41
Late Glacial Maximum (LGM), 45, 49, 108
Late Holocene period, 89; climate summary, 101–2; geomorphological evidence, 95–96; isotopic data, 89–91; lacustrine evidence, 92–95; pollen evidence, 91–92
Late Natufians, 111, 113, 123, 124–27
Late Pleistocene environment, 67–69

leaders. *See* ruling elite
Lisan, Lake, 64–69
lowland economy, 138–39

Magalit Terrace, 61–63
maps: Byzantine settlements, 152; Dead Sea/Lake Lisan, 66; Early Bronze Age, 133; Natufian sites, 110; Tel 'Erani site, 131
Mediterranean Sea, 90, 166–67
Mediterranean zone, modern, 32–33
medium-scale variability, 4–6
Middle Bronze Age (MB), 40–41
Middle Holocene period, 80; climate summary, 99–101; geomorphological evidence, 86–88; isotopic data, 81–84; lacustrine evidence, 85–86
mobility of Natufians, 106–7, 111
monocropping, 140–41

Nabateans, 160–61
Natufian period, 34–36, 45, 59–60
Natufians: agricultural origins and, 104–7; buffering devices, 121; compared to other hunter-gatherers, 118–25; Early society of, 109–11, 122, 124, 174; Late society of, 111, 113, 123, 124–27; map/structures, 110, 112
Negev Desert, 160; economic prosperity, 161–64; Holocene proxy data, 79, 82–83, 89; Roman-Byzantine proxy data, 165–68; terrace systems, 63–64, 87–88
Neolithic sites, 36–37, 78–79
Norse responses to stress, 11, 148
North American hunter-gatherers, 120–22, 124–25

oak (*Quercus*) pollen analyses, 55–61, 75–76, 84–85, 92
oasis theory, 1, 103–4
olives (*Olea*): economic role, 137–38, 140–41; pollen analyses, 84–85, 91–92, 132
oxygen isotopic data, 25–27; consistency of, 44–45; Early Holocene, 71–73, 74; eastern